Strategic Environmental Performance

Sustainable Improvements in Environment Safety and Health

Series Editor

Frances Alston

Lawrence Livermore National Laboratory, California, USA

Lean Implementation
Applications and Hidden Costs

Frances Alston

Safety Culture and High-Risk Environments
A Leadership Perspective

Cindy L. Caldwell

Industrial Hygiene
Improving Worker Health through an Operational Risk Approach

Frances Alston, Emily J. Millikin, and Willie Piispanen

The Legal Aspects of Industrial Hygiene and Safety
Kurt W. Dreger

Strategic Environmental Performance
Obtaining and Sustaining Compliance

Alston Frances and Brian Perkins

For more information about this series, please visit: https://www.crcpress.com/Sustainable-Improvements-in-Environment-Safety-and-Health/book-series/CRCSUSIMPENVSAF

Strategic Environmental Performance

Obtaining and Sustaining Compliance

Dr. Frances Alston and
Brian K. Perkins

CRC Press
Taylor & Francis Group
Boca Raton London New York

CRC Press is an imprint of the
Taylor & Francis Group, an **informa** business

First edition published 2020
by CRC Press
6000 Broken Sound Parkway NW, Suite 300, Boca Raton, FL 33487-2742

and by CRC Press
2 Park Square, Milton Park, Abingdon, Oxon, OX14 4RN

© 2021 Taylor & Francis Group, LLC

CRC Press is an imprint of Taylor & Francis Group, LLC

ISBN: 9780367471552 (hbk)
ISBN: 9781003038894 (ebk)

Typeset in Times
by Deanta Global Publishing Services, Chennai, India

Contents

Authors

Dr. Frances Alston has built a solid career leading the development and management of Environment, Safety, Health and Quality (ESH&Q) programs in diverse cultural environments. Throughout her career, she has delivered superior performance within complex, multi-stakeholder situations and has effectively dealt with challenging safety, operational, programmatic, regulatory and environmental issues.

She has been effective in facilitating integration of ESH&Q programs and policies as a core business function while leading a staff of business, scientific, engineering and technical professionals. Dr. Alston also has extensive knowledge and experience in assessing programs and cultures to determine areas for improvement and development of strategies that facilitate positive outcomes.

Dr. Alston holds a B.S. degree in industrial hygiene and safety, an M.S. degree in hazardous and waste materials management/environmental engineering, an M.S.E. in systems engineering/engineering management and a Ph.D. in industrial and systems engineering.

She is a Fellow of the American Society for Engineering Management (ASEM) and hold certifications as a Certified Hazardous Materials Manager (CHMM) and a Certified Professional Engineering Manager (CPEM). She is Adjutant Professor at Oregon State University and the 2018 President for the American Society for Engineering Manager.

Brian K. Perkins has more than 30 years of experience in environmental safety and health, quality assurance, radioactive waste management and environmental remediation. He has held various management positions in a private industry and state government and has supported U.S. Department of Energy activities at the Nevada National Security Site, the Yucca Mountain Project and Lawrence Livermore National Laboratory.

Mr. Perkins has successfully led multidisciplinary teams in the development and implementation of efficient and sustainable environmental compliance and waste management programs in various complex and highly regulated arenas. His program management and operational experience also includes environmental management systems, integrated safety management, nuclear facility safety and emergency response.

He earned his B.S. degree in geology from the University of Nevada Reno, an M.S. degree in environmental management from the University of San Francisco and has held certifications as an Environmental Manager, Environmental Auditor, NQA-1 Auditor and Hazardous Material Manager.

1 Environmental Compliance and the Corporate Structure

1.1 DEFINING THE COMPLIANCE PROFILE

No matter what industry or the type of business an organization is engaged in, establishment of a corporate compliance structure is a must. A compliance structure or profile simply refers to identifying the laws and regulations that must be adhered to and the road map for ensuring compliance with applicable laws. Compliance is becoming more synonymous with conducting business no matter the industry or market in which a company competes. Regulatory compliance can range from those that apply to people management, from the viewpoint of human resource management, safety and health of workers while performing work, to protection of the environment. Recognizing that compliance for each company can be different depending on the mission of that company, the question that is important for each organization to contemplate is, what is compliance for my company? Compliance, simply put, is a company's ability to identify, conform to and follow the rules and regulations that apply to them as they conduct business. The applicable rules and regulations form the basis of a company's compliance profile. However, for some companies, it is not always easy to identify all the regulations that govern the ways in which they conduct business. In such cases, some organizations discover through regulatory inspections that they are not in compliance with a law or regulation that they have not even considered as a part of their compliance profile for the business. The compliance profile of a company is dependent upon various variables such as:

- The number of workers that a company employs
- The types of hazards that an employee can be exposed to during the performance of their tasks
- The type of potential emissions and releases to the environment
- The type of products produced
- The types of chemical products utilized while conducting business
- The type of waste or by-products produced from conducting business
- The state in which business is being conducted and the regulatory requirements specific to business conducted within that state

In addition to paying attention to the variables that can impact the structure of compliance, corporate leaders must also focus on the prevailing top compliance issues as they unfold. Keeping abreast of these issues provide an opportunity for a company to examine or assess compliance and the mechanisms used to address any of the applicable prevailing issues. These issues may include:

- Complexities encountered when managing risks for the enterprise
- Understanding and reducing the potential of regulatory noncompliance (Managing environmental risks falls within this category.)
- Managing changes in the information technology arena to include reducing the potential of cyber infiltration and attacks
- Developing and managing an effective internal audit program

Total compliance for a company means that they are conforming to *all* of the environmental regulations and laws. These laws and regulations can also be documented in permits or agreements that the company may have with the regulatory agencies such as Consent Orders (CO). To facilitate compliance, many organizations make use of environmental checklists, solicit the assistance of workers having the most knowledge of the operations that they are responsible for supporting, and hire environmental engineers, specialists or professionals who have been trained in environmental regulations. Environmental laws are legally enforceable primarily through the Environmental Protection Agency (EPA). However, the EPA does not typically respond to all environmental concerns as some environmental issues are handled by other federal, state or local agencies through delegated authorization agreements from the EPA. The EPA has in many cases provided enforcement responsibility to some state and local governmental agencies. When enforcement is delegated, these agencies have the authority to inspect and impose penalties for noncompliance that may include financial penalties or imprisonment for failure to comply with the laws and regulations. The cost of noncompliance for a company can also lead to increased inspections by regulatory agencies, tarnish a company reputation and instill the perception of being an organization that is willing to cut cost and quality and having little or no concern for workers and the environment.

1.2   ALIGNMENT WITH THE MISSION OF THE COMPANY

When considering alignment of a company's mission and compliance profile, two key questions come to mind. The answer to these questions can be important when it comes to gaining support from the leadership team and workers in supporting activities that will facilitate compliance even if they believe that some actions taken may not be viewed as significant in supporting business practices and perhaps even hindering productivity. Why is it necessary for the compliance profile of a company to be in alignment with the mission of the company? How important is it for this alignment to take place? The mission of a company gives direction and purpose to that company. A well-documented mission statement is short and defines the goal, customer base and the products or services they will provide. Once the mission of the company is flushed out, its leadership must ensure that the organization is structured in operable units that will effectively support the mission, create cohesion and assist the organization in reaching its peak performance. The term peak performance is defined as the point at which an organization has implemented streamlined processes that has a positive impact on the business, reduces rework, the amount of waste generated and disposed of, human error is anticipated and managed, and employees are engaged in the business (Allen, Alston, Millikin Dekerchove, 2019). Organization systems are recognized as a system that is connected to assisting in the completion of tasks and activities that support achievement of the goal and mission of an entity. As such, when developing these systems, it is important to take a strategic view and ensure that the system used is in alignment with the mission of the company.

The mission has a great bearing on the type of laws that are applicable to the business and the compliance posture that a company must assume. Therefore, environmental compliance is not feasible unless it is strategically aligned with the mission of the company. For example, Company Y is in the business of producing a solvent that yields waste by-products that are released through a stack into the environment. In order to operate the process, the company will need to submit an air permit application to obtain an air permit. The permit and its content are legally binding and must be followed just as they would for any environmental regulation. If the process did not emit organic solvent vapors into the air, an air permit would not have been necessary and the compliance requirements that come with an air permit would not have been necessary. The company must also ensure compliance with waste and other applicable regulations. Consider another scenario for Company X whose mission is to produce a product that is considered relatively nonhazardous, producing no emissions to the air, with some waste generated from the by-products. Company X typically will not be required to obtain and implement an air permit, although they must comply with waste regulations. The point here is that environmental compliance considerations should begin at the forefront of business development, and the compliance posture is significantly dependent upon the business practice and its potential impact on people and the environment as well as the state in which the business is performed.

1.3 ORGANIZATIONAL STRUCTURES THAT ENABLE COMPLIANCE

Organizations today face unprecedented levels of complexity having to operate in dynamic environments with a knowledge-based culture and fast-paced economy. Many customers are seeking products that are unique and companies that provide exceptional service and quality, and they are no longer satisfied with what may be considered as standard products. The knowledge age is forcing leaders to focus on coordinating resources in a way that requires managing knowledge capacities as a means to compete. Mobilizing knowledge resources provides an organization with the ability to provide high-value services and deliver novel solutions to solve problems (Yoo, Boland, Lyytinen, 2006). These knowledge workers are critical in the development of exceptional services and non-standard products.

There are different ways that an organization can be organized into an effective functioning system that supports both the needs of the organization and inclusiveness among workers and leadership. In an organization, many rules and regulations must be applied and followed by all to ensure compliance. These regulations include those that are applicable to, for example, financial matters, human resource and environment safety and health. Recognizing that all structures have the ability to enable compliance to some degree, there are some structures that can better enable and support environmental compliance.

Organizational structure gives members a clear guideline on roles and responsibilities and how to proceed with getting work accomplished, and it can also serve as a means to bind members together as they work to accomplish a common goal. The CEO or the president has the responsibility to structure the organization so that it can handle all of the problems they will face and provides a support path to conduct business effectively and with the maximum level of productivity feasible. In structuring the organization, considerations must be given to the issues that the organization may face in the future and to a recognition that changes in one part of an organization will impact other areas of the organization. Positioning an organization for success requires continual effort in analyzing internal efficiency and external uncertainties that may be posed in the future (Cyert, 1994). There are four types of organization systems that are used fluently today: functional, divisional, matrix and flat systems. These systems are discussed in further detail in Sections 1.3.1–1.3.4.

1.3.1 FUNCTIONAL ORGANIZATION STRUCTURE

Functional organization systems structures group people based on their area of specialization and assign leadership to those having expertise in the same area. The grouping of people is placed in units referred to as departments with a department manager, who is held accountable for performance of the department. Effective utilization of workers is expected since the manager understands from a technical standpoint the tasks performed by workers. Having managers with technical understanding is expected to help the organization achieve some level of effectiveness in obtaining its objectives due to their knowledge and familiarity with the tasks that need to be performed. An example of a functional system is shown in Figure 1.1. Each unit is responsible for handling one aspect of ensuring success of the product or services provided, for example human resources, engineering, infrastructure and services, information technology, environmental compliance, etc.

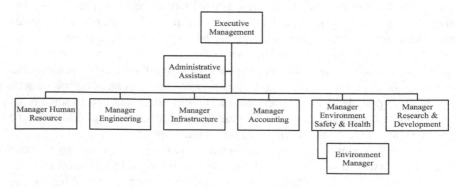

FIGURE 1.1 Functional organization structure.

TABLE 1.1

Benefits and Pitfalls of Functional Systems

Potential Benefits	Potential Pitfalls
Assist in controlling consistency of quality and performance	The departments often function as silos disconnected from the rest of the company
Having all authority and decision-making, including budget and resource allocation, remains with the functional manager	Communication flows vertically through department heads to the president or CEO
Minimal time and funds spent on training as tasks and business functions change infrequently	Employees may become distracted and lose their questioning attitude as a result of performing the same tasks over time
Employees report to one manager	Compartmentalized knowledge with little opportunities to expand knowledge and opportunities beyond skills needed to support the department
No duplication of responsibility and efforts	The interdepartmental communication can cause employees to lose touch for what is happening in the company as a whole and may not understand how the work they perform contribute to the overall mission

The functional organization structure is effective for organizations that have ongoing operations and produces standardized products. Some benefits and pitfalls of a functional system are listed in Table 1.1.

The environment manager reporting to the environment safety and health manager (ES&H) is a common reporting line and provides direct access to the rest of the safety and health functions as there are times when these functions have to work closely together to ensure that environmental policies are implemented effectively. However, regardless of the reporting structure, the group that has direct responsibility to facilitate and enable compliance should report at a high level within the organization and should have direct access to the senior leadership team and the president or CEO. Some of the benefits and pitfalls that may be encountered in flowing down environmental requirements and facilitating compliance in a functional organization are documented in Table 1.2.

The functional system may be a great choice of ease in implementing an environmental compliance program within a company; however, in today's environment, it may not be the structure of choice for organizations that have a variety of business lines or for organizations that are interested in research and development and other evolving technological ventures.

1.3.2 DIVISIONAL ORGANIZATION STRUCTURE

A divisional organization system is designed by splitting up an organization into semiautonomous areas referred to as division. Each division consists of a complete set of functions that are essential for achieving goals, as shown in Figure 1.2. The

TABLE 1.2

Flow Down of Environmental Regulatory Requirements in a Functional System

Potential Benefits	Potential Pitfalls
Groups requiring knowledge in environmental compliance are easily identified	Employees may become distracted and lose their questioning attitude as a result of performing the same tasks over time. Therefore, they may not keep abreast of environmental regulatory requirements or report issues to management as they occur
Flow down of environmental regulatory requirements simplified due to the departmentalized structure	Less eyes and perhaps attention to issues that can lead to environmental noncompliances
Fewer people to train and track qualification requirements	
Supervision accountability easily to identify and assess	
Smaller number of people and area involved in assessment of environmental matters	

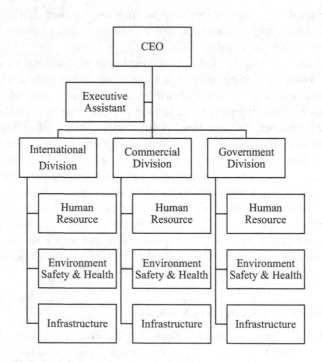

FIGURE 1.2   Divisional organization structure.

divisional system makes use of the functional system and groups it into divisions to support the product or market in which they compete. The divisional organizational system is extremely beneficial for companies that conduct business in different regions, markets or product lines.

The challenges and benefits involved in developing and implementing an environmental compliant program in a divisional structure are listed in Tables 1.3 and 1.4. Although not a complete list, the tables provides some issues that should be considered and leveraged or avoided in some cases.

1.3.3 MATRIXED ORGANIZATION STRUCTURE

A matrix organizational structure combines two or more types of organizational structures. In this structure, employees have more than one boss and may be expected to support several projects at a time. Leadership responsibilities are often split between the functional manager and the project manager. This type of structure makes use of the concept of 'dual authority' and generally combines the divisional and functional structure in forming the matrix system (Harris and Raviv, 2002). A matrix system allows sharing of resources across functions, which is an ideal use of the matrix organization structure. An example of a matrix organization system is shown in Figure 1.3. When considering the four primary structures used most

---

TABLE 1.3

Benefits and Pitfalls of Divisional Systems

Benefits	Pitfalls
Each division is able to focus efforts on a specific product, service or market in which they compete	Higher operational cost for the company is expected because duplication of resources exists as each division is fully autonomous having all resources necessary for supporting each division or group
Facilitate business decisions and the coordination of activities among divisions to be made expeditiously by the local leadership team because they have complete autonomy for organizational matters	Each division will have its own strategic focus than may not completely support the goals of the company as a whole
Facilitation of a common culture within each division	The company may not be able to take advantages of economies of scales because purchases of goods are not typically integrated across the company
Development of knowledge and expertise in the products and services	Communication across the company is minimized
Places decision-making as close to the customer as possible	Operate in a silo with little or no collaboration with other division employees within the company
Can respond faster to changes in the market as decision-making is localized	

TABLE 1.4

Flow Down of Environmental Regulatory Requirements in a Divisional System

Potential Benefits	Potential Pitfalls
Quick flow down of information and changes in environmental regulatory requirements	Regulatory flow down may be inconsistent due to several environmental departments operating in silos
Common compliance culture within each division	Potential differences in regulatory interpretation across the company and flow down information
	The perception or acknowledgment of having too many resources across the company and risk reduction in force during critical budget times
	Compliance culture different across the company
	Environmental compliance costly across the company due to duplication of resources and efforts

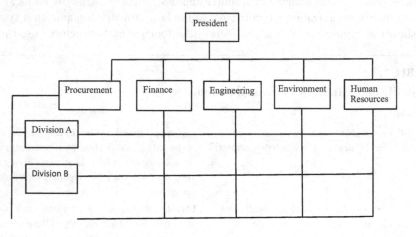

FIGURE 1.3   Matrix organization structure.

frequently today, the matrix system presents more challenges when it comes to developing and implementing a compliant regulatory system and culture.

The matrix system presents many complexities when it comes to flow down and implementation of environmental regulations because roles and responsibilities across the organization may not be clear (Tables 1.5 and 1.6). In this type of system, it is often difficult to keep track of projects that have environmental law impacts because of the different layers of groups that must be communicated with and the movement of personnel to support the various projects within the organization.

Developing a strategy to support environmental compliance in the matrixed organization presents complexities that involve having many levels of groups of people performing projects and tasks that involve compliance considerations that must be overcome to ensure compliance with laws and regulations.

TABLE 1.5
Some Benefits and Pitfalls of Matrix Systems

Potential Benefits	Potential Pitfalls
Sharing of knowledge and skills across the company	Responsibility for employee development may not be clear to employees
Fosters employee development	Heavy workload for employees because of the support provided for several projects or departments
Cost less due to the sharing of resources	Handling conflicts may be complicated because many layers of management may be involved due to reporting structure
Efficient use of limited resources	Source allocation and assignment issues when many projects are in need of the same skill sets
Project structure assistance in development of resources and encourage team cohesion	Employees are accountable to more than one manager with at least two chains of command

TABLE 1.6
Flow Down of Environmental Regulatory Requirements in a Matrix System

Potential Benefits	Potential Pitfalls
Several groups or teams to hold accountable for compliance	Difficulty in ensuring new workers have the training and knowledge needed to adhere to regulatory requirements
Many eyes on environmental compliance	Flow down of regulations may not reach workers actually impacted by regulatory requirements
Cost of compliance less due to sharing of resources	The need to train many workers in various aspects of environmental regulatory requirements
	Tracking regulatory training requiring and compliance is complicated due to the number of workers trained
	Management accountability since it may be difficult to determine process and task responsibility for managers and hold them accountable
	More areas requiring inspections

1.3.4 FLAT ORGANIZATION STRUCTURE

A flat organizational structure is an organization system that does not have layers of management between the company's grassroots staff members and senior managers. In these structures, employees usually report to the president, the CEO or the highest level of leadership instead of reporting to a lower level of management that then reports to the senior executive. Flat organization structures are not common for medium- to large-size organizations. This type of structure is typically effectively utilized in small or start-up companies. An example of a flat organization structure is shown in Figure 1.4.

Environmental compliance in small organizations tends to be less complex and easier to achieve because of the absence of organizational layers of management and

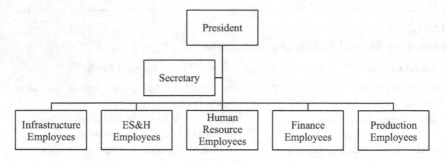

FIGURE 1.4 Typical flat organization structure.

also because these organizations tend to have small groups of employees. The ES&H or the environmental compliance manager and subject matter experts directly report to the top level of management and have fewer organizations to consult with and enable compliance to regulations and laws. In addition, it is possible that since these organizations tend to be small, the environmental compliance posture will include a limited number of laws and regulations, depending upon the business profile and functions. In other words, the simpler the structure and business functions, the easier the compliance (Tables 1.7 and 1.8).

TABLE 1.7
Benefits and Pitfalls of Flat Systems

Potential Benefits	Potential Pitfall
Cost efficient, having fewer layers of management	Employee retention may be difficult due to lack of job change opportunities and promotions
Clear and consistent communication. Miss communication is minimized	Management can lose control if organization expands having a large ratio of employee to manager
Expedited decision-making	

TABLE 1.8
Flow Down of Environmental Regulatory Requirements in a Flat System

Potential Benefits	Potential Pitfalls
Easier flow down and implementation of environmental requirements since requirements is expected to be less for a small organization with a smaller business portfolio	The perception that compliance poster is insignificant
Fewer employees needing training and retraining	Minimal staffing to address compliance issues
Less regulatory inspections from regulators	Less environmental knowledgeable staff for regulatory interpretation and flow down
Less internal inspections to verify compliance	

No matter what organization system is utilized to organize a company, environmental compliance must be at the forefront and not an afterthought. When environmental compliance is planned as an afterthought, one can expect difficulty in compliance and thus potentially expect fines to be levied when not in compliance with the applicable laws and regulations.

1.4 APPLIED LEARNING

Review the case study and respond to the questions that follow.

1.4.1 CASE STUDY

The senior leadership team in a major international company having a matrix organization system decided that a process expansion be made to keep up with and improve production of a proprietary component that will change the capability and use of a chemical product. The changes in production will include additional chemicals that are not a part of the current process. To support the new process, the company will have to hire additional resources.

1. Should the company reevaluate their compliance posture? If yes, describe the process or strategy that will be used to determine the new compliance posture for the company. If no, why should the company continue business as usual?
2. Is the process change significant enough that the leadership team need to reevaluate the organization structure to support the process changes? Explain your decision.
3. How should flow down of the new applicable environmental requirements be handled?
4. How will you address training and knowledge requirements to the new staff and current staff who will be responsible for new process operation?
5. What impact will the current organization structure exert on the process expansion?

REFERENCES

Allen, Patricia Melton, Alston, Frances E., & Dekerchove, Emily Millikin. (2019). *Peak performance: how to achieve and sustain excellence in operations management.* Boca Raton, FL: CRC Press/Taylor & Francis.

Cyert, Richard. (1994). Positioning the organization. *The Institute of Management Sciences, 24*(2), 101–104.

Harris, Milton & Raviv, Artur. (2002). Organization design to organization designing. *Management Science, 48*(7), 853–865.

Yoo, Youngjin, Boland, Richard J. Jr., & Lyytinen, Kalle. (2006). From organization design. *Organization Science, 17*(2), 215–229.

2 Corporate Culture

2.1 INTRODUCTION

Culture within organizations has been studied for decades and continues to gain importance as organizations are positioning themselves to perform at their maximum potential and provide value for their customers. Many authors and theorists have studied and contributed to documented research into the role culture plays in organizational success. The preponderance of these studies suggests that culture can be created to exhibit the creator's values, priorities and vision. These studies further suggest that culture is key to employee commitment, productivity and corporate profitability. One of the most notable recognized contributors to cultural research is Edgar Schein, who composed the most popular definition of organizational culture. He defines culture as

> a pattern of shared basic assumptions that the group learned as it solved its problems of external adaptation and internal integration that has worked well enough to be considered valid and, therefore, to be taught to new members as the correct way to perceive, think and feel in relation to those problems.

Important components of a great culture include a vision and a mission that are supported by members of the organization; clear, sound values that are communicated via documents serving as guidelines for expected behaviors; sound practices that demonstrate value for their employees and customers and a leadership team that 'walk the talk'; and employees who are engaged. A company's culture can be a viable source of sustainable competitive advantage because culture determines behaviors that bond and motivate organizational members, and the shared values and beliefs that are present within a culture are important variables that guide behaviors (Alston 2014).

A study of the relationship between the culture of an organization and the level of trust demonstrated that the two concepts are related. The results specifically showed that employees' perception of their work culture and trust in their organizations are related; therefore, this aspect must be in the forefront of all leadership strategies. It is also important that considerations be given to actions that can lead to mistrust. Frequent triggers of mistrust include the abrasive and abusive conduct as well as ambiguous behaviors of management and supervisors and the employees' perception about the organization's culture. It is easier to achieve trust when an organization's aim, vision, mission, values, objectives and goals are understood and shared.

The culture of an organization serves as the foundation of the organization and enhances the ability of members to build relationships and trust among members. Human performance is significantly dependent on the culture since culture has an influence on members in terms of releasing creativity and productivity (Alston 2014).

2.2   TOP-DOWN LEADERSHIP: DEMONSTRATED COMMITMENT

A highly successful company has employees who are committed to it and its mission. These employees have an unwavering dedication that is evident in the pride they have in what they do and the organization in which they perform their roles. High levels of organizational commitment are associated with organizational attributes such as high levels of performance seen through increased productivity, employee retention and increased profit margins. Organizational commitment is being broadly defined as the degree of commitment and loyalty that employees have for their employer. To achieve these benefits and more, many organizations engage in and focus on improvements in the following areas:

Creating and Nurturing a Strong Culture: Building a strong culture that takes advantage of the diversity that each employee brings to the workplace establishes a bond between workers and an opportunity to learn as they begin to understand each other and the value of their differences to the organization. Promoting a culture of diversity and inclusion provides the motivation needed for people to work together as they begin to understand and respect each other and their contribution to the team.

Focus and Model Trusting Behaviors: Trust is viewed as a central component of a strong work culture. When employees have trust in organizational members, they are willing to sign on to the values and goals of the organization. In addition, they are more willing to exert the attention to what is needed to accomplish tasks. Management behavior is a key component of achieving and retaining trust.

Recognize and Celebrate Successes: Celebrate often the success of both the individual employees and the organization. The act of celebrating and rewarding individuals for exceptional performance demonstrates value for their contribution to the success of the mission and builds confidence in themselves and the work they do.

Facilitate Employee Engagement: Engaged employees are destined to be more involved with and committed to the business, and in most case they will remain with the company for an appreciable amount of time. Engaged employees are instrumental in helping organizations discover cost-effective ways of performing work, developing new technologies and products and expanding services for customers.

Ensure Goals and Expectations Are Clearly Communicated: Communication of goals and expectations is a must so that employees can perform to a desired level and assume accountability and ownership for performance. When employees feel a sense of ownership, they tend to become more engaged and remain with the organization longer.

Transparency in Communication and Actions: Transparency in communication and actions increases trust, builds relationships and increases creativity, engagement and productivity. Transparency in organizations means that employees have access to unfiltered information about the company to include performance data.

Understand Humans Are Fallible and Mistakes Will Be Made: Because humans are fallible, there are times when mistakes will be made. When mistakes occur, use these occurrences as teachable moments. Evaluate the mistake and identify the human element that may have resulted from inadequate process or procedure and rectify to ensure mistakes are not repeated.

Encourage Strategic Thinking and Innovation: Encouraging and allowing employees to utilize their creativity is one of the best ways to encourage employees to become engaged in business activities.

2.2.1 LEADERSHIP COMMITMENT

Implementation of the mission and goal of an organization is an important responsibility of leadership. In order to accomplish this role, a leader must be committed to the organization and its vision. Leadership commitment is verbally communicated with the accompanying actions. For the most part, employees will become committed to the organization easily if they believe that the leadership team is committed. How do leaders demonstrate their commitment to the organization? Some actions are listed below:

- They demonstrate concerns for employees.
- They develop and implement policies and procedures that are in the best interest of the business, taking into consideration the impact of those policies and procedures on the workers and their ability to perform.
- They walk the talk. Leadership follows the policies and procedures that they expect of others.
- They conduct themselves in a manner that they are considered trustworthy.
- They lead with honesty and humility.

Employee commitment is being defined as an individual having a strong connection for their job, the organization and to other members of the organization. This connection is displayed in the actions and attitudes of the individual. Commitment can be viewed as a bond that holds an individual to their organization and the vision set forth by leadership. A key thing to remember is that there will be no employee commitment without leadership-demonstrated commitment.

2.3 EMPLOYEE BUY-IN AND ENGAGEMENT

Buy-in from employees should be sought after with the realization that not every employee will buy into and support every initiative or decision made on behalf of the organization. Employee buy-in is a great precursor to engagement. When employees accept the decisions of leadership, they are more likely to go the extra mile to accomplish work.

Employee engagement has many recognizable benefits for organizations to include:

- Increased productivity and the desire to assist the organization to achieve at their highest level and meet the needs of their customers
- Increased facilitation of employee growth
- Empowers employees to participate in the business
- Increased development and maintaining of trust
- Facilitate employee retention

- Increased creative and strategic thinking
- Better decision-making that is in alignment with the goal and vision of the organization

In order for engagement to be highly effective as a strategy, not only must employees be engaged in their organizations, they must also believe that they are included as an integral entity in helping achieve goals, objectives and the mission (Alston 2014).

2.4 CULTURAL IMPACTS ON COMPLIANCE

Leadership has the responsibility of creating a culture of compliance for their organizations. The first step in culture creation involves the actions of the leadership team and the expectations set for organization members. Leaders must demonstrate honesty and transparency in dealing with implementing regulatory requirements and responding to inquiries concerning their compliance posture. The culture of an organization drives behaviors of the employees as they perform their job responsibilities. According to Alston 2014, based upon an extensive literature review, there are four primary cultural attributes that are important in defining and sustaining the culture of an organization. These attributes, namely, practices, values, beliefs and behavior, play a critical role in developing and maintaining a compliance culture. Each of these attributes will be discussed in Sections 2.4.1–2.4.4.

2.4.1 THE BELIEF ATTRIBUTE

Substantial influence can be placed on an organization by the beliefs that are present within an organization because the engrained beliefs represent important variables that guide behaviors. This influence can and generally will have impacts on how to operate in a compliant manner regardless of the requirements imposed by the prevailing regulations. If the leadership team demonstrate their belief in a compliant culture, then the expectation is that their subordinates will adopt that same belief. Because beliefs are ideas and ideals that people hold as being factual, compliance can be impacted when inaccurate beliefs are acted upon. For example, a belief that compliance documentation is not necessary to be retained for future reference can equate to documentation is not always kept. Failure to produce documents required to be viewed by a regulator during an inspection will lead to a noncompliance and potentially an imposed financial penalty.

2.4.2 THE BEHAVIOR ATTRIBUTE

Human behavior can be destructive or constructive and can reinforce the need for compliance or discourage workers from performing work in a compliant way. There are three aspects of human behavior that can significantly impact compliance and culture:

- People tend to follow leaders that they have confidence in.
- People seek unity and tend to follow others to avoid peer pressure.
- People react and respond to pressure.

Considering these aspects as they are operating in a compliance culture, leaders must demonstrate their support and expectations for employees to perform work according to procedures, policies and regulatory requirements.

2.4.3 THE PRACTICES ATTRIBUTE

Practices are the unwritten activities and ways of conducting business based on past rules or on ways of thinking. These practices are ingrained by workers who have been with the organization for a long period of time. These ingrained ways of thinking and conducting business are extremely difficult to change in many cases, and these practices are handed down to new workers when they enter the organization. Although newly assigned workers are aware of the desired way of conducting business, they may succumb to the practices that are performed due to peer pressure or the desire to fit in and not be singled out as someone having the desire to be different.

Ingrained practices can have a significant impact on the implementation of environmental rules and regulations. Just as environmental laws and regulations change, implementation must also change to complement the required intent to meet compliance. For example, effective from January 1, 2014, the EPA revised the reporting requirements for the chemical inventory that a company has on their site (https://www w.epa.gov/assessing-and-managing-chemicals-under-tsca/tsca-work-plan-chemic al-assessments-2014-update). The new requirements included the need to collect additional data that were not previously required. This new requirement requires workers to be trained and keep the revised reporting requirements in mind or perhaps consult the regulations before collecting facility information as the old practice and data collection needs have changed.

2.4.4 THE VALUES ATTRIBUTE

Values refer to the standards that are routinely deemed as acceptable or unacceptable in an organization. Cultural values directly impact human interactions and behaviors that will have a direct impact on the propensity to place compliance needs on the same playing field as getting work accomplished. Organizational members must place value in ensuring that the organization develop and follow procedures and practices that will ensure compliance to all requirements. Shared values consist of the goals that are shared by people in an organization that shape and guide behaviors. The behaviors exhibited by people that are directed toward following policies and procedures have an impact on compliance.

2.5 EVALUATING CULTURE FOR EFFECTIVENESS

There are several methods that can be used to measure and evaluate the cultural health of an organization. These methods include reviewing documents such as policies and procedures, administering a survey instrument, interviewing individual employees and conducting both focus group discussions and work observations. Collectively, these methods have provided useful information that can be used to determine the health of culture.

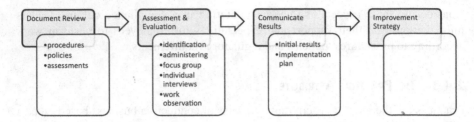

FIGURE 2.1   Culture evaluation process.

The literature on culture states, in general terms, that culture can be created to display the creator's values, priorities and vision. Developing and nurturing a culture that can facilitate business growth can be challenging due to the diversity of people and the different customs they bring to organizations and development of varying subcultures that underline the overall culture of an organization. The process that has been effectively used in evaluating culture is shown in Figure 2.1.

Documentation Review: Perhaps the first step in evaluating the culture of an organization should start with a review of applicable policies, procedures and charters that document the expectations for the way that employees are expected to conduct business and interact with each other. In organizations where the policies and procedures are clearly documented and communicated, workers are most likely to support making every attempt to comply with the expectations set by these documents. When there is a lack of documentation that clearly supports the vision of the organization, workers can become confused and unsure of the appropriate behavior and expectations. This ambiguity will have some impact on the compliance posture of an organization. Conducting a review of documentation as the first step of the process can produce important information that can inform the survey, focus group and individual interview process.

Assessment and Evaluation: Many organizations use several mechanisms to conduct a comprehensive survey of their work culture. These methods include the use of survey instruments, documentation review, work observation, focus group and individual interviews. When using surveys to assess culture, there is a need to ensure that the proper survey instrument is selected. There are many instruments on the market today that are used to measure culture; therefore, there are many to choose from. When using developed instruments, ensure the following:

- The survey instrument contains the questions needed to provide answers that complement the purpose of the evaluation and the feedback being sought.
- The survey has been validated.
- The questions are clear and not written in a way that can be perceived as leading.

Some potential constraints that must be considered when utilizing survey methods are as follows:

- The method or methods used to analyze the data.
- The cost associated with procuring and administering the survey instrument.
- How will the survey be administered? Paper distribution or electronic.
- The time required to complete the survey.
- Does the survey ask the right questions?

Conducting focus groups and individual interviews is another effective way of gathering data to support a comprehensive and effective culture evaluation. A focus group should consist of 6–12 participants. The reason for keeping the group small is to facilitate free flow of information exchanges. When utilized appropriately, focus group discussions are a valuable tool used to discover more about the opinions of employees. Guidelines to consider when conducting focus group discussions include:

- Keep the list of questions short to no more than 15 questions.
- Use terms that are familiar to the participants.
- Use open-ended conversation-type questions.
- Do not use leading questions.
- Each question should focus on a single element or activity.
- Select participants randomly from a group that has knowledge of the topic.
- Select a meeting location that would ensure privacy for the group.

Individual interviews can be conducted when there is a need to gather information from one individual. This method is an effective way to get feedback from senior managers. The same questions should be addressed for each focus group discussion. In addition to the questions asked during the focus group discussions, additional questions that may be appropriate only for senior managers may be asked during those interviews. Individual interviews should be kept to the prescheduled time frame. The individual and focus group interviews are good processes to collect additional data on the cultural health of an organization. The information obtained will provide insight into the views and opinions of the leadership team and the workers (Allen, Patricia, Alston, Frances, Dekerchove Emily, 2019).

Work observations provide an opportunity for the evaluators to observe work and see how work is actually performed by workers. Observing work also provides an indication of whether the policies and procedures are followed and work is paused if there is a lack of understanding of procedure or if a regulatory noncompliance exists or a safety hazard that needs to be addressed. The actions of workers during work performance provides clues to the workplace safety culture. Some actions to pay attention to during work performance include:

- Was a pre-job briefing performed before work began?
- Did the workers who will perform the job participate in the pre-job briefing?
- Was the scope of the work discussed and followed?
- Did the worker perform the work according to procedure, work documentations, etc.?
- Were safe work practices used?

- Were the tools needed to perform work on the job site available to the workers?

Communication: When embarking upon the culture evaluation process, it is paramount that a communication plan be developed and followed. Communicating to workers the need to perform a culture assessment and the process that will be used to conduct the data collection is important. Once the culture assessment is completed, a comprehensive communication campaign should be conducted to provide employees the result of the assessment and management's commitment to address any weakness in the organization culture. The improvement strategy is communicated to workers as soon as it is developed, outlining the improvement areas and the process for addressing those areas.

Improvement Strategy: Many companies are beginning to understand the value and importance of including in their overall business strategy provisions to ensure that the culture of the organization is evaluated periodically and changes are made as necessary. A comprehensive culture improvement strategy should be completed once the data obtained from the culture assessment have been analyzed and areas of improvement have been identified. The plan should address all areas of weakness identified. It is not necessary to address all issues at the same time, however; all issues should be included in a schedule with a time frame for addressing them. Metrics should be developed to track the plan progress and made available for all employees to view.

2.6   APPLIED LEARNING

1. What impact can the culture of an organization have on environmental compliance?
2. List and explain the potential impacts of the four key cultural attributes on regulatory compliance in organizations. Discuss the role of leadership in compliance.
3. Read the two scenarios below and discuss the culture described and its impact on compliance, the attitude and perceptions of leadership and employees.

2.6.1   SCENARIO 1

On Monday, the security guard from A Company telephoned the Environment Safety & Health office that representatives from the Environmental Protection Agency were at the site gate and seeking access to the facility. The director and the president of the company agreed that they were not ready for a no notice inspection and so requested that the security guard stall three team regulators for about an hour to give them the opportunity to conduct a quick inspection of some problem areas of the plant. The employees were tasked with quickly inspecting their work spaces and remove any noncompliant conditions, concealing them in any manner possible.

After about an hour and a half, the inspectors were allowed to enter the facility and inspect the five preselected areas. The inspectors identified the following noncompliances:

- Waste was not being handled and disposed of properly.
- Chemicals were stored improperly.
- Documentation demonstrating maintenance of equipment designed to facilitate compliance was not available.
- Workers were not forthcoming in responding to questions.

The regulatory inspection report left behind by the inspectors was not embraced by leadership, and employees were requested to correct the violations because management wanted to avoid being asked any fines because of impacts on the quarterly profit margin and on managers' bonuses.

2.6.2 SCENARIO 2

On Monday, the security guard from B Company telephoned the Environment Safety & Health office and informed that representatives from the Environmental Protection Agency were at the site gate and were seeking access to the facility. The director and the president of the company agreed that the security guard should grant the regulators immediate access to the facility. The employees were alerted that the inspectors would be in their work area and the expectation was that they support the inspections in any way they could. They were further informed that any noncompliant conditions should be rectified as soon as feasible.

The inspectors identified the following noncompliances:

- Waste was not being handled and disposed of properly.
- Chemicals were stored improperly.
- Documentation demonstrating maintenance of equipment designed to facilitate compliance was not available.

The regulatory inspection report left behind by the inspectors was reviewed carefully by management and a plan put in place to resolve each issue and prevent reoccurrence.

REFERENCES

Allen, Patricia, Alston, Frances, & Dekerchove, Emily. (2019). *Peak performance how to achieve and sustain operations management.* Boca Raton, FL: CRC Press/Taylor & Francis Group.
Alston, Frances. (2014). *Culture and trust in technology-driven organizations.* CRC Press Taylor/Francis Group.
https://www.epa.gov/assessing-and-managing-chemicals-under-tsca/tsca-work-plan-chemic al-assessments-2014-update

3 Identifying Applicable Requirements

3.1 ENVIRONMENTAL REGULATIONS OVERVIEW

Environmental regulations in today's world constitute a highly complex set of requirements that can be overwhelming when it comes to understanding applicability and identifying mechanisms and processes for implementing effective compliance programs. Environmental professionals tasked with environmental compliance must possess a thorough understanding of the operations and processes taking place within their company or organization in order to understand which environmental regulations apply. Considering the enormous number of different processes and/or operations occurring within business and industry each day, it becomes evident that there is no simple, one-size-fits all program for environmental compliance. Compliance programs for large corporations, small businesses, manufacturing processes, service industries and government at the federal, state and local municipality levels, all must be tailored to fit each company or organization. What this requires is a strategic approach that ensures applicable regulatory requirements are properly identified, processes are established that ensure consistent and compliant implementation and roles and responsibilities are clearly defined and understood.

Laws targeting environmental protection were basically nonexistent during the industrial revolution and have only become an ongoing focus of lawmakers over the last half century. There are environmental-related acts or laws dating back to the early 1900s. These include such laws as the Lacey Act and the first passage of the Federal Insecticide, Fungicide and Rodenticide Act (FIFRA). The Lacey Act was passed in 1900 as an environmental conservation measure to protect wildlife and plants from illegal trade through the enforcement of civil and criminal penalties (Randy Browning 2019). FIFRA was enacted in 1910 to protect farmers and people consuming agricultural products from deceptive pesticide manufacturers distributing sham pesticide products and/or pesticide products that contained levels of arsenic and lead that were dangerous to livestock and human health (Ganzel 2019). Both laws are still in place today although they have both been substantially modified through numerous amendments and legislative actions.

A growing environmentalism movement that began in the 1960s and continued into the 1970s was spawned by a series of environmentally related events and failed environmental policies. Incidents such as the discovery of the deleterious effects of pesticides (DDT) on waterfowl and other birds as documented in Rachael Carsen's book *Silent Spring*, the 1969 oil spill of the coast of Santa Barbara, California, the Cuyahoga river catching fire in 1969 and the increasing occurrence of acid rain, all fueled an emergent wave of environmentalism that swept the country during the 1970s. In response to the growing environmentalism movement, the Nixon Administration began an unprecedented campaign to develop environmental-based legislation focused on the protection of the environment and the natural resources. One of the most notable actions occurring under the Nixon Administration was the signing of the National Environmental Policy Act (NEPA), which included the creation of the Council on Environmental Quality (CEQ). The CEQ was developed to advise the president on environmental issues and review environmental impacts from federal agency projects. Recommendations from the CEQ would translate into the establishment of the Environmental Protection Agency (EPA) in 1970.

Many of the primary environmental laws that are in place today were enacted during the Nixon Administration and throughout the 1970s. Most of the original environmental laws enacted during the 1970s have been amended, numerous times, but the purpose and intent of their development is founded in the environmentalist era of the 1970s. Major legislation such as the Clean Air Act (CAA), which was originally enacted in 1963 as a measure to evaluate air pollution and research air pollution control measures, was amended under the newly formed EPA to authorize the development of standards for air emissions for industrial plants and for mobile sources of air emissions. The need for controlling air emissions from mobile sources such as automobiles had been recognized years before when Congress passed the National Emissions Standard Act in 1965 to require automobile manufacturers to lower pollution from automobiles. The Clean Water Act (CWA) is another fundamental environmental legislation that was originated during this period. Developed to protect water resources and clean up polluted waters, the CWA, much like the CAA, has evolved into a very comprehensive and foundational set of regulations that led the way in environmental protection legislation.

Other key environmental legislation enacted during the 1970s include: the Endangered Species Act (ESA), which was set up to provide protection for species of plants and wildlife that were in danger of becoming extinct; the Safe Drinking Water Act (SWDA), which set quality standards for drinking water; the Resource Conservation and Recovery Act (RCRA), which created a cradle-to-grave regulatory system for managing the generation, treatment, storage, transportation and disposal of hazardous waste; the Toxic Substances Control Act (TSCA), which set regulatory standards for the development and manufacturing of synthetic chemicals; and the Federal Environmental Pesticide Control Act, which amended earlier versions FIFRA regulations for controlling the manufacture and use of pesticides, rodenticides and fungicides. This era of environmentalism during the 1970s marked an unprecedented number of legislative actions focused on protection of the environment and human health.

3.2   REGULATORY REQUIREMENTS

Since the explosion of environmental legislation and rapid promulgation of environmental protection regulations during the 1970s, the world population has grown from approximately 3.6 billion in 1970 to over 7.6 billion people in 2019, and in the United States alone the population has grown from 205 million to an estimated 330 million during this same time span (US Department of Commerce 2019). This exponential population growth places ever-growing demands for basic necessities such as food and clean water and other consumer goods and services. These growing demands ultimately equate to an unprecedented reliance on critical natural resources. The industrial world and commerce are continually evolving to meet these growing demands with new and innovative technologies aimed at improving efficiency and increased production. The impacts of population growth and the resulting demand for increased production of consumer goods and services require that environmental protection legislation and associated regulations at the federal, state and local levels progressively move forward in order to ensure ongoing protection of human health and the environment. What this equates to is a very dynamic and complex regulatory setting for environmental compliance and with such a vast number of regulations, managing compliance can be overwhelming. Determining which regulations apply to your business or operation becomes a critical element in developing an effective environmental compliance program. The following sections discuss the major environmental laws in place today, associated regulations, their purpose and scope and implications associated with compliance. In addition, the discussion includes practical approaches and proven methods that can be used in evaluating regulations and associated requirements to determine their applicability to your business or operations.

3.3 FEDERAL LAWS

3.3.1 THE CLEAN WATER ACT

The origins of the Clean Water Act date back to 1948 when the Federal Water Pollution Control Act (FWPCA) was first enacted. The original legislation was passed by Congress as a means to address concerns regarding increasing pollution of lakes and rivers resulting from the industrial expansion stemming from World War II. The FWPCA of 1948 turned out to be an ineffective legislation and did very little to address the ongoing and unmitigated discharge of pollutants and a growing water pollution problem. Subsequent amendments to the FWPCA during the 1950s and 1960s were made to expand the role of the government, but enforcement of the law was difficult and there was very little progress in cleaning up polluted waters (Copeland 2016). The emerging environmentalism movement in the 1970s was fueled by a heightened level of concern regarding the degree of pollution in lakes and waterways and a growing level of public awareness on just how severe the problem had become. Under the newly formed EPA, the FWPCA underwent a major revision in 1972 that gave EPA the authority and established the framework for regulating discharges of pollutants into waters of the United States and setting standards for industrial wastewaters. Included in the 1972 amendments were the establishment of laws making it illegal to discharge pollutants into waters from a point source without first obtaining a permit and the authorization of grant program that would provide funding for the construction of sewage treatment plants (U.S. EPA 2019). Since the major revisions in the 1972 amendments, the FWPCA, most commonly referred to as the CWA, has undergone amendments in 1977, 1981, 1987 and again in 2014.

3.3.2 THE CLEAN WATER ACT TODAY

Today the CWA consists of provisions which authorize federal assistance in wastewater treatment plant construction and requirements applicable to industrial and municipal wastewaters dischargers. The federal assistance programs for wastewater treatment plants have evolved programs at the state level, where states match federal funding for the construction of wastewater treatment plants and are refunded monies used for these purposes under a revolving loan-type program. The regulatory requirements of the CWA are the aspects related to facility compliance and will be the focus of discussion here.

The regulatory framework of the CWA covers the following six primary elements: wastewater management, pretreatment, storm water pollution, animal waste from concentrated feeding operations, spills of oils and hazardous substances and wetlands protection. The EPA regulations for water programs are found in Subchapter D, Chapter 40 of the Code of Federal Regulations in Parts 100–149 (40 CFR 100–149). The following is a summary discussion of the aforementioned six primary elements of the CWA.

3.3.2.1 Wastewater Management

Wastewater management regulations focus on the discharge of pollutants from industrial and municipal wastewater collection and treatment plants and storm water discharges from both municipalities and industrial operations. Industrial facilities and municipalities must obtain permits under the National Pollutant Discharge Elimination System (NPDES) prior to discharging wastewater or allowing storm water to enter waters of the United States. NPDES was an important element added to the CWA in the 1972 amendment and remains a focal part of the regulations. EPA has delegated implementation of the NDES program to states and tribal governments with established programs that have been authorized and approved by the EPA. Massachusetts, New Hampshire, New Mexico and the District of Columbia are the only states that currently do not have their own NPDES programs, and the EPA remains the lead agency for issuing NPDES permits in these states. This is also the case for the U.S. territories and most tribal lands.

There are two types of permits under NPDES: an individual permit, which is for an individual facility where the controls identified in the permit are developed based on the operations, processes, pollutants being discharged from that specific facility, and the waters receiving the discharge. The second type of NPDES permit is a general permit that is intended to cover a group of facilities with similar discharges located within the same geographical area.

The fundamental aspect of an NPDES permit is to set limits on the amounts and types of pollutants that can be discharged from a point source into the waters receiving the effluent. In this way, the quality of the receiving waters can be protected. The limitations specified within an NPDES permit are based on available technology for controlling pollutants and the water quality standards that have been set for the water receiving the discharge. The CWA also established a process for states to identify waters within their jurisdiction where water quality standards cannot be met solely by technology-based treatment. In these cases, states develop total maximum daily loads (TMDLs) for these priority waters, and these TMDLs become effluent limitations in NPDES permits. TMDLs are developed for each priority water and identify the pollutants and specific amounts that may be discharged from all sources (point, nonpoint and natural) while ensuring that the water quality standards for that specific water body are maintained.

Two important terms used within the NPDES program that are pivotal for determining applicability and whether an NPDES permit is needed are 'point source' and 'waters of the U.S.' The EPA defines point source as *any discernable, confined, and discrete conveyance, including but not limited to, any pipe, ditch, channel, tunnel, conduit, well, discrete fissure, container, rolling stock, concentrated animal feeding operation, landfill leachate collection system, vessel or other floating craft from which pollutants are or may be discharged. This term does not include return flows from irrigated agricultural storm water runoff* (40 CFR, Part 122.2 2019). Based on this rather broad definition, the task of determining whether a discharge is a 'point source' can be fairly simple, but it is often times not a straightforward determination. The second aspect of evaluating the need for an NPDES permit is the determination of whether an effluent is being discharged into waters of the United States. Waters of the United States can be summarized as navigable waters and their tributaries, interstate waters (including wetlands), intrastate waters used for recreation and commerce and the oceans out to 200 miles. The actual definition is found in 40 CFR Part 122.2 and is much more detailed and complex than the summary above as it also includes any waters adjacent to those described above and any waters that may have a significant nexus (significant chemical, physical or biological effects) to any of the waters summarized above. The definition of the waters of the United States has been a point of ongoing legal challenges for decades due the inherent complexity and subjectivity involved in applying the definition. Understanding NPDES permitting requirements is a critical element of a compliance program for any operations involving wastewater and effluent discharges in order to avoid potential regulatory noncompliance and the possibility of fines and penalties for illegally discharging without a permit. Under the authority of the CWA, regulatory agencies authorized to enforce CWA requirements can levy fines and penalties ranging from up to one year in jail and $2,500–25,000 per day for negligent violations. The penalties for knowingly discharging pollutants to waters of the United States can include up to three years in jail and $5,000–50,000 per day per violation. For both negligent and knowing violations, these penalties can be doubled for subsequent convictions.

Wastewater discharges from industrial and/or commercial operations into surface waters, impoundments, storm drains or directly to the ground surface most likely will require an NPDES permit. It is important to understand that the term 'pollutant' is not limited to only chemical wastes; it also includes solid waste, incinerator residue, filter backwash, sewage, sewage sludge, garbage, munitions, biological materials and radioactive materials. The regulatory status of these types of discharges should be discussed with the regulatory agency having authority for NPDES prior to actively discharging. As previously mentioned, EPA has delegated authority for NPDES to state environmental protection agencies in most cases.

Once an NPDES permit is issued, it will contain permit requirements for effluent limitations, periodic testing of the effluent and reporting of the test results to the regulatory agency. The periodic testing will require that samples of the effluent be collected and submitted for laboratory analysis utilizing specified analytical methods. The permit conditions for effluent sampling and reporting requirements will vary based on the characteristics of the discharge and the waters receiving the effluent.

3.3.2.2 Pretreatment Program

NPDES permits are not required for wastewater that is discharged directly into municipal sanitary sewer systems [publicly owned treatment works (POTW)], but discharges to sewers may require a permit from the municipality itself. POTWs issue discharge permits in order to control the types and amounts of pollutants being discharged from nondomestic sources to the sanitary sewer and eventually the treatment plant. POTWs control discharges through permitted releases to ensure that the operation of the sewage treatment plant is not affected by toxic or incompatible pollutants that would interfere with or disrupt operations. These permits also help ensure that pollutants aren't being discharged into the POTW that could be released to the environment by 'passing through' the treatment plant or impact the management and disposal of sewage sludge (U.S. EPA 2019).

The pretreatment program is a national program that is a coordinated effort between EPA, the states and local municipalities. Under this program, municipal POTWs have been granted the authority by the EPA and authorized states for the permitting and enforcement aspects related to discharges into their sanitary sewer systems and POTWs. The pretreatment program identifies discharge standards for nondomestic wastewaters discharged into POTWs. The discharge standards include general and specific prohibitions, categorical pretreatment standards and local limits as needed. The general prohibitions forbid the discharge of any pollutant that will pass directly through the POTW at a concentration exceeding what the POTW is allowed to discharge or any discharge that would interfere or disrupt the treatment processes. Specific prohibitions forbid any discharge that is corrosive (<5 pH), would cause a fire or explosion, obstruct flow, is more than 104°F or would cause the generation of toxic, fumes, vapors or gases. The categorical pretreatment standards are effluent limitations that apply to specific industrial categories such as electronics manufacturing, electroplating operations, explosives manufacturing and cement manufacturing, to name a few. The general and specific prohibitions and the categorical pretreatment standards are all national standards that apply to all POTWs and nondomestic dischargers regardless of whether they have been issued a permit. POTWs may develop local limits that are specific to the operations of the POTW and the types of discharges being received by the POTW (U.S. EPA 2019).

Similar to an NPDES permit, a permit issued by a POTW to a nondomestic discharger will include permit conditions and requirements that identify discharge limitations, requirements for sampling the effluent to demonstrate effluent limitations are being met and regulatory reporting requirements.

3.3.2.3 Storm Water Pollution

Another important component of the NPDES program is the storm water pollution program. The storm water program regulations are focused on controlling storm water that can potentially entrain or pick up pollutants as it flows across impervious surfaces and transport these contaminants into rivers, lakes, streams and/or coastal waters. The potential for storm water to pick up and transport pollutants is much higher at sites conducting industrial activities and at constructions sites. As such, the storm water program regulates activities from three distinct types of sites: municipalities, constructions sites and industrial activities. Activities occurring at these types of sites may need to obtain an NPDES permit before allowing the discharge of storm water. Similar to wastewater discharges from point sources and the pretreatment program for POTWs, EPA has delegated the authority to implement the storm water program to the states (U.S. EPA 2019). Controlling storm water pollution is accomplished by the use of what is termed 'Best Management Practices' (BMPs) within the regulations. BMPs include controlling pollution at the source by properly storing chemicals and materials, covering soil/debris piles to protect from storms, expeditious cleanup of spills, use of filters and other media to filter out pollutants, use of vegetation cover to control erosion and limit storm water runoff and maintaining areas with surfaces subject to storm water flow so they are clean and free of pollutants. For operations/activities requiring an NPDES storm water permit, a Storm Water Pollution Prevention Plan (SWPPP) must be prepared. A SWPPP is a regulatory document that describes how storm water will be controlled at the site, which BMPs will be implemented, methods of implementation and other requirements such as site inspections, frequency of storm water sampling, analytical parameters for specific pollutants and regulatory reporting (U.S. EPA 2019).

3.3.2.4 Construction Storm Water

Construction projects that disturb 1 acre or more will typically require a storm water permit. Recall from the earlier discussion on NPDES permits that there are two types of NPDES permits: individual and general. Construction storm water permits are covered under construction general permits (CGP). States with authorized storm water programs have a state CGP, and there is an EPA CGP for those areas that are not covered by state and tribal programs. Construction projects needing a storm water permit must apply for coverage under either their state CGP or the EPA CGP depending on which agency has jurisdiction over the area. Figure 3.1 shows a decision flowchart developed by the EPA that is intended to help in determining if an NPDES storm water permit is needed. CGPs are comprehensive documents that include information and requirements covering the following main topical areas: how to obtain coverage under the CGP; technology-based and water quality-based effluent limitations; site inspection requirements; requirements for corrective actions; training requirements, SWPPP requirements and how to terminate coverage under the CGP. The CGP will also contain the requirements for certifying that specific

**Do I need to get covered under an NPDES Construction General Permit (CGP)
for stormwater discharges for my construction site?**

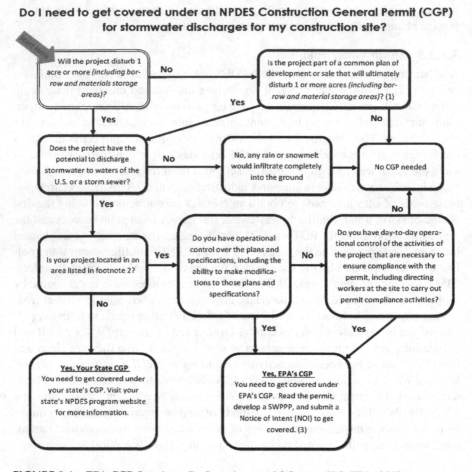

FIGURE 3.1 EPA CGP flowchart: Do I need a permit? Source: U.S. EPA, 2017.

criteria have been met regarding threatened and/or endangered species and compliance under the ESA. There are also National Historic Preservation Act (NHPA) requirements related to screening for historic properties. These are some of the main requirement areas found in CGPs; however, there are additional requirements found in CGPs, and there are differences among the various state CGPs and the EPA CGP. Therefore, it is important to understand which agency has jurisdiction over the area the construction activity is occurring in and to understand the requirements of that agency's CGP. The EPA and most states have their CGPs available on their websites, where they are readily accessible.

3.3.2.5 Industrial Storm Water

Specific industrial activities are also subject to storm water NPDES permitting requirements. EPA defines *storm water discharge associated with industrial activity* as: 'the discharge from any conveyance that is used for collecting and conveying storm water that is directly related to manufacturing, processing or raw materials storage areas at an industrial plant.' (U.S. EPA 2019) The specific industrial activities are defined in 40 CFR 122.26(b)–14(i) and include the following:

1. Facilities subject to federal storm water effluent limitation guidelines, new source performance standards and/or toxic pollutant effluent standards (with some exceptions) standards per 40 CFR 405–471
2. Heavy manufacturing (e.g., chemical plants, petroleum refineries, steel mills)
3. Mining, oil and gas exploration and processing
4. Hazardous waste treatment, storage and disposal facilities
5. Landfills and open dumps with industrial wastes
6. Metal scrapyards, salvage yards, automobile junk yards and battery reclaimers
7. Steam electric power generating plants
8. Transportation facilities with vehicle maintenance activities, equipment cleaning of airport deicing operations
9. Facilities treating domestic sewage with a daily flow through of ≥ 1 million gallons per day
10. Constructions sites that disturb ≥5 acres
11. Light manufacturing such as food processing, printing and publishing, electronic and electrical equipment manufacturing, public warehousing and storage.

Similar to storm water permits for construction activities, industrial storm water permits are often issued under general permits. In areas where EPA has jurisdiction, storm water discharges from regulated industrial activities are covered under the Multi-Sector General Permit (MSGP), which is titled the *NPDES Multi-Sector General Permit for Stormwater Discharges Associated with Industrial Activity* (U.S. EPA 2019). Most states authorized to implement the NPDES program also utilize general permits for regulating storm water discharges from industrial activities. Some states utilize industrial general permits that cover regulated industrial activities under a single permit. As an example, the state of Nevada utilizes an MSGP for all industrial activities with the exception of mining, which is covered under a general permit specific to mining operations (NDEP 2019), whereas other states such as Alabama have numerous general permits, with each general permit covering specific industrial categories (e.g., asphalt manufacturing, metal finishing/fabrication, textile industry, landfill operations) (ADEM 2019).

3.3.2.6 Municipal Storm Water

Cities, towns, and other urban areas own and operate systems designed to capture and transport storm water as a means to prevent flooding and control the collection and conveyance of storm water. These storm drain systems are known as municipal separate storm sewer systems (MS4s), and these systems by definition do not discharge into a POTW. As with construction and industrial sites, MS4s have the potential to collect and convey storm water that has pickup up pollutants as it has flowed across impervious surfaces. Polluted storm water entering an MS4 can then be transported within the storm drain system and potentially discharged into waters of the United States without being treated. States, cities, towns and counties that own MS4s are required to obtain NPDES permits to cover discharges of storm waters from these MS4s. Operators of MS4s are required to develop, implement and enforce a storm water management program (SWMP) that details how the MS4 will control and reduce the discharge of pollutants from the system. Based on EPA guidance, SWPPs should include discussion on how the MS4 will reduce pollution from construction site runoff, detect and eliminate illicit discharges, pollution prevention and good housekeeping practices, controlling postconstruction runoff and programs for public participation, education and outreach, program effectiveness and total maximum daily loads (U.S. EPA 2019).

3.3.2.7   Concentrated Animal Feeding Operations

Animal feeding operations where animals are kept confined on a lot or within a facility can contribute pollutants such as nitrogen, phosphorous, pathogens, hormones, antibiotics and other organic wastes and wastewater and/or storm water. Animal feeding operations that meet specific size requirements as defined by EPA in 40 CFR 122.23 are considered concentrated animal feeding operations (CAFO). The EPA defines a CAFO based on the number and type of animals that are involved in the feeding operation. Under the NPDES program, CAFOs are considered point sources and require an NPDES discharge permit (U.S. EPA 2019). Similar to NPDES wastewater discharge permits, NPDES permits for CAFOs will contain requirements for effluent limitations, periodic testing of the effluent and reporting of the results of the testing to the regulatory agency.

3.3.2.8   Spills of Oils and Hazardous Substances (SPCC)

In 1990, the CWA was amended by the Oil Pollution Act to require facilities that use or store certain quantities of oil to prepare plans for responding to oil discharges, and the plans must address a worst case spill or discharge scenario. Under the oil pollution prevention regulations (40 CFR Part 112), nontransportation-related facilities must develop SPCC plans to prevent oil discharges. Facilities that store larger amounts of oil may also be required to develop a facility response plan (FRP). A facility may need to develop an SPCC plan if it is a nontransportation facility that has a total capacity of more than 1,320 gallons of oil in aboveground storage or a capacity of more than 42,000 gallons of oil in completely buried storage tanks. Oil-filled equipment with tanks or other containers used to store oil solely to support the function of the equipment or apparatus (e.g., hydraulic presses, transformers, electrical switches, gear boxes) must be included when determining total capacity. It is important to note that storage containers that have less than 55 gallons capacity and underground storage tanks (USTs) that are regulated under the UST regulations in 40 CFR 280 or 281 are not included when determining total oil storage capacity for SPCC. The types of oil that are covered under the SPCC regulations are not limited to petroleum oil, but they also include oils, fats, greases from animals and fish, vegetable oils, oils from seeds and nuts, petroleum fuel oil, sludge, synthetic oils, mineral oils and waste oils. A 'non-transportation' facility is basically any industrial, commercial agricultural or public facility that stores or uses oil, whereas a 'transportation facility' is one with operations involving the movement of oil from one facility to another.

The other critical aspect of whether a facility is subject to the SPCC requirements is the geographical location of the facility. This criterion is based on whether a facility is located in an area where it could reasonably be expected to discharge an amount of oil that may be harmful into navigable waters of the United States or adjoining shorelines. The terms *may be harmful* and *navigable waters* are important to understand here as they are pivotal in determining applicability. The term *may be harmful* is defined in 40 CFR Part 110.3 and basically means any amount of oil that will cause a sheen or film on the water; create a sludge or emulsion; or violate water quality standards applicable to the affected water body. While there is no actual quantity or volume of oil associated with the term, the quantity of oil required to cause a sheen or film on water can be a very small amount. The term *navigable waters* means waters of the United States as defined by the CWA (U.S. EPA 2019). As discussed in the preceding section on wastewater, waters of the United States is a very broad term that includes not only actual navigable waters and their tributaries but also any adjacent waters, wetlands and any waters that may have a significant nexus (significant chemical, physical or biological effects) to any of these waters.

Nontransportation facilities that meet the oil storage capacity thresholds discussed above and are geographically located such that there is a reasonable expectation that oil could be discharged into navigable waters are required to develop an SPCC plan. Figure 3.2 is an SPCC applicability flowchart developed by EPA that is intended to aid in determining if your facility needs to develop an SPCC plan. For facilities that are required to develop an SPCC plan, the regulations identify three different types of facilities for the purposes of defining the required content of an SPCC plan and whether the plan can be 'self-certified' or must be certified by a professional engineer (PE).

3.3.2.9 Tier I Facility

Facilities that have less than 10,000 gallons total aboveground storage with no individual container greater than 5,000 gallons that have not had a single discharge of more than 1,000 gallons into navigable waters within three years of certifying the plan or two discharges exceeding 42 gallons each within a 12-month period qualify as a Tier I facility. Operators of a Tier I facility have the option to develop an SPCC plan following the template found in 40 CFR Part 112 Appendix G. This option provides for a less detailed plan and can be self-certified by the owner/operator of the facility. When self-certifying an SPCC plan, owners/operators are attesting that they are familiar with the facility and have visited and evaluated the facility. Owner/operators are also certifying that the plan has been prepared in accordance with requirements and acceptable industry standards, the facility meets the Tier I criteria, the plan is being implemented fully, there are no deviations from the regulatory requirements, the plan has been approved by management and the resources needed to implement the plan have been made available (U.S. EPA 2019).

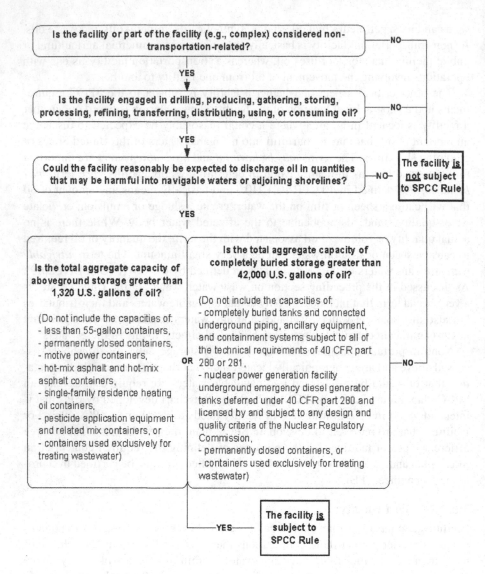

FIGURE 3.2 EPA SPCC applicability flowchart. Source: U.S. EPA 2017.

3.3.2.10 Tier II Facility

A facility that meets the requirements described above for a Tier I facility with the exception that there are individual aboveground storage containers with capacities greater than 5,000 gallons qualifies as a Tier II facility. Tier II facilities are required to develop an SPCC plan in accordance with the general requirements found in 40 CFR 112.7 and meet the requirements in subparts B and C. The general requirements for an SPCC plan found in 40 CFR 112.7 describe the required content and sequence for the plan. Subpart B of 40 CFR 112 lays out the requirements for SPCC plans relative to facilities storing and/or using petroleum oils and nonpetroleum oils except animal fats, greases and oils and oils derived from vegetables, nuts, seeds and kernels. 40 CFR 112 Subpart C in turn contains the SPCC plan requirements for facilities using and/or storing animal fats, greases and oils and oils derived from vegetables, nuts, seeds and kernels. Similar to a Tier I facility SPCC plan, owners/operators of a Tier II-qualified facility can also self-certify their SPCC plan (U.S. EPA 2019).

Facilities that are subject to the SPCC requirements but that do not qualify as a Tier I or Tier II facility must have their SPCC plan certified by a PE. Owners/operators of these facilities must meet the requirements of 40 CFR 112.7 and subparts B and C much like Tier II facilities with the difference being that the SPCC requires certification by a PE for the initial plan and for any technical amendments to the approved plan.

SPCC regulations are intended to prevent discharges of oil to navigable waters. All SPCC plans (Tier I, Tier II and larger facilities) must address the following common elements:

- Facility diagram and description of facility
- Oil discharge predictions
- Sized secondary containment for all bulk oil storage containers (designed to hold the full capacity of the container plus an additional volume for precipitation that may occur)
- Facility drainage
- Facility security
- Ongoing periodic inspections and/or testing of tanks, piping and documenting the inspections, including any deficiencies, repairs or other identified issues
- Provide for overfill prevention via high-level alarms or 'spill buckets' to catch overflows
- Procedures for containment of potential spills during transfers of oil into and out of containers (mobile refueling, product transfers)
- Using appropriate type of containers for the product being stored
- Personnel training
- Current and accurate inventory of all SPCC-regulated containers (bulk storage containers, oil-filled equipment, mobile refuelers, etc.)
- Five-year review of the SPCC plan and/or periodic updates for technical changes requiring an amendment to the SPCC plan
- Management approval

SPCC plans are required to be maintained at the facility and do not require submission to the EPA unless requested. Facilities that are not staffed more than four hours per day must maintain the SPCC plan at the nearest office or business location (U.S. EPA 2010).

3.3.2.11 Facility Response Plan

Facilities storing larger quantities of oil may be required to develop an FRP. These are facilities that could be reasonably expected to discharge oil to navigable waters and cause substantial harm. The FRP rule defines a facility that could pose substantial harm as one having:

- A total oil storage capacity ≥42,000 gallons and transfers oil over water to/from vessels, or
- A total oil storage capacity ≥1,000,000 gallons and meets one of the following:
  - Does not have sufficient secondary containment for aboveground storage areas
  - Is located such that a discharge could cause injury to fish, wildlife or sensitive environments
  - Is located such that a discharge would shut down a public drinking water intake
  - Had a reportable discharge of ≥10,000 gallons within the past five years.

Under the FRP rule, facility owners/operators determine if they meet the above criteria and if their facility is subject to the FRP requirements. Under the FRP rule, the above criteria are termed the 'self-identification criteria.' Facilities that are determined to meet the self-identification criteria are subject to the FRP rule and must prepare an FRP and submit it to their regional EPA office. FRPs that are submitted to EPA regional offices are evaluated by EPA to determine if the facility could cause 'significant and substantial harm.' When performing this evaluation, EPA considers the substantial harm criteria described above plus other factors such as age of storage tanks, types of transfer operations, storage capacity, lack of secondary containment, facility location relative to sensitive environments, fish, wildlife, drinking water intakes, facility spill history and any other relative information. Facilities that are determined by the EPA to have the potential to cause 'significant and substantial harm' must have their FRP reviewed and approved by the EPA (U.S. EPA 2018).

The primary components required for an FRP include the following:

- Facility information (name, location, owner, facility type)
- Facility site diagrams, drainage, evacuation plan
- Description of site security
- Emergency response action plan
- Information regarding emergency notification, equipment, personnel and evacuation
- Analysis of potential spills and identification of previous spills
- Description of scenarios for various size spills (small to worst case) and associated response actions
- Procedures for discharge detection and associated equipment
- Detailed plan for spill response, containment and disposal
- Records and documentation related to inspections, drills/exercises and response training

In addition, the facility owners/operators must identify in their FRP the qualified individual who is fully authorized to implement corrective and/or removal actions and communicate with the federal authorities and responders. The owner/operator must also identify the resources to respond to and remove a worst case discharge and ensure the availability of those resources. Any significant changes to an FRP must be submitted to the EPA regional office for review and approval. An example of an FRP is provided in Appendix F of 40 CFR 112 to assist owners/operators in preparing an FRP (U.S. EPA 2018).

3.3.2.12 Wetlands Protection

Section 404 of the CWA regulates the discharge of dredged or fill materials into waters of the United States, which includes wetlands. The types of activities regulated under Section 404 include such activities as the use of fill materials for development projects, construction of dams, levees, airports, highways and mining-related activities. Section 404 of the CWA requires that an activity must be permitted prior to dredged material or fill being discharged to waters of the United States. The Section 404 permitting process strives to minimize impacts to surface waters and wetlands by requiring that less damaging alternatives have been adequately evaluated, potential impacts are minimized and that any impacts that are unavoidable are mitigated through compensation.

Compensation to offset unavoidable impacts can be through the restoration, creation, enhancement or preservation of aquatic resources. Restoration involves reestablishing or rehabilitating physical, chemical or biological characteristics of a resource area so that it functions the way it used to, or as close as practical to its historical function. Compensation can also involve the creation of new aquatic resources that didn't previously exist, the enhancement of specific functions of an aquatic resource or the preservation of an aquatic resource to prevent a decline in its function (U.S. EPA 2019).

These compensatory mitigations are realized through one of three different processes. The first is through utilization of a mitigation bank to buy mitigation credits. Buying mitigation credits basically is a means by which a permittee is paying to transfer the obligation of compensatory mitigation to the owner/sponsor of the mitigation bank. This mechanism allows for the permittee to buy the appropriate number of credits (commonly based on acreage or area of impact) from a mitigation bank that owns a site or multiple sites where aquatic resources are restored, created, enhanced or preserved. Buying credits from a mitigation bank is basically paying for mitigation that has already taken place (i.e., in the bank) as a means to compensate for unavoidable impacts to aquatic resources caused by an activity or project. The second mechanism is very similar to a mitigation bank is called an 'In-lieu fee program.' Much like a mitigation bank, the In-lieu fee program sells compensatory mitigation credits to permittees as a means to compensate for impacts on aquatic resources. The In-lieu fee program also involves sites where aquatic resources are restored, created, enhanced or preserved, but these are accomplished by nonprofit or government agencies that receive funding specifically for the purpose of mitigating aquatic resources and making compensatory mitigation credits available to permittees. The third mechanism is where the permittee maintains the responsibility for restoring, creating, enhancing or preserving an aquatic resource (U.S. EPA 2019).

Both the U.S. EPA and the U.S. Army Corps of Engineers (USACOE) have roles under Section 404 of the CWA. The USACOE is responsible for reviewing and issuing permits, developing guidance, day-to-day administration of Section 404 permit program and for enforcing Section 404 permitting requirements. The USACOE also conducts and verifies jurisdictional determinations. These are a pivotal aspect of permit program as these determinations identify what are considered to be 'waters of the U.S.' and in turn potentially subject to Section 404 of the CWA.

Section 404 permits issued by the U.S. ACOE can be either individual permits or general permits. General permit can be issued on a nationwide, regional or state basis. General permits are beneficial in expediting the issuance of permits by eliminating or minimizing delays in permitting process when a proposed activity meets the conditions for a general permit. Examples of activities that can be covered under a general permit are backfilling utility trenches and minor road work/repairs.

The U.S. EPA has responsibilities related to developing policy, guidance and environmental-related criteria to be considered when evaluating permit. The U.S. EPA also enforces Section 404 requirements, oversees state and tribal program implementation and reviews individual permit applications (U.S. EPA 2019).

3.3.3 THE CLEAN AIR ACT

The Clean Air Act was originally established in 1963 in response to increasing smog and air pollution in many American cities and industrialized areas. The CAA of 1963 was focused more on studying air pollution and did very little to mitigate or clean up air pollution. The basic framework for the CAA that is in place today was established by Congress in 1970 during the 'environmentalism movement,' as discussed earlier in Section 3.2. The establishment of the CAA was in response to growing concerns over smog and air pollution that had continued to worsen in the absence of any real pollution prevention or cleanup efforts. The increasing emission of air pollutants was associated with a noted increase in respiratory issues and other deleterious effects on both human health and the environment. The CAA of 1970 established programs at the federal and state levels for limiting emissions from industrial sources as well as mobile sources. The programs established in the CAA of 1970 were the National Ambient Air Quality Standards (NAAQS), New Source Performance Standards (NSPS) and the National Emission Standards for Hazardous Air Pollutants (NESHAPs). The act also established requirements for states to develop implementation plans detailing how these new air pollution control regulations were to be implemented at the state level. These state implementation plans (SIPs) laid the groundwork for EPA authorizing states to regulate and enforce regulations promulgated under the CAA.

The CAA was amended in 1977 to add requirements to prevent deterioration of air quality in areas that were meeting the air quality standards (attainment areas) defined in NAAQS. The 1977 CAA amendments also added requirements for areas that were not meeting one or more of the federal air quality standards under NAAQS (nonattainment areas).

The CAA underwent a major set of amendments in 1990 that expanded the NAAQS and NESHAPs requirements and established new programs to address acid rain and ozone. The 1990 CAA amendments also expanded the federal government's enforcement authority and programs for air pollution control research (U.S. EPA 2017).

3.3.4 THE CLEAN AIR ACT TODAY

The CAA regulations today comprise a complex and thorough collection of regulations and requirements that have been promulgated by the EPA as authorized under the CAA legislation and amendments. Much like the CWA, EPA has delegated authority to the states for implementing and enforcing CAA regulations. The CAA is organized into six main areas, which are referred to as Titles I through VI. The following discussion will provide a summary of the major requirements under each of the CAA Titles and is not intended to a complete and comprehensive discussion.

3.3.4.1 Title I

Title I of the CAA is subdivided into the following four parts:

- Part A – Air Quality and Emission Limitations
- Part B – Ozone Protection (replaced by Title VI in the 1990 CAA amendments)
- Part C – Prevention of Significant Deterioration of Air Quality
- Part D – Plan Requirements for Nonattainment Areas

Part A included requirements for states to develop implementation plans (SIPs) which describe how primary and secondary ambient air quality standards will be met and maintained. An integral part of the SIP process was the requirement for states to identify air quality control regions for the purpose of managing air emissions and attaining federal air quality standards.

Under Title I of the CAA, EPA established NAAQS for sulfur oxides, carbon monoxide, particulate matter [includes particulate matter ≤ 10 micrometers (PM_{10}) and particulate matter ≤ 2.5 micrometers ($PM_{2.5}$)], nitrogen oxides, lead and ozone. These six pollutants are known as *criteria* air pollutants. The NAAQS identifies both primary and secondary air quality standards. Primary standards define the levels necessary to protect human health and especially sensitive populations such as children, older adults and those with respiratory-related health problems. The secondary standards were established to provide protection to animals, crops and other vegetation and protection against visibility issues associated with smog (U.S. EPA 2016). Air quality control regions that have air quality that meets or is better than the primary ambient air quality standards as defined by EPA are called *attainment areas*, whereas regional areas not meeting the primary ambient air quality standards are known as *nonattainment areas*. Attainment areas are regulated in a manner such that the air quality is maintained and primary ambient air quality standards are not exceeded. Nonattainment areas are required to develop plans for reducing air emissions with the goal of attaining air quality that meets the primary standards for the criteria pollutants. Obtaining air permits in nonattainment areas can be more restrictive than in attainment areas, and permit conditions to operate are commonly more rigorous due to the state or local regulatory agency's obligation to reduce emissions of criteria air pollutants in the area. The regulations pertaining to NAAQS and SIPs are found in 40 CFR Parts 50 and 51, respectively.

Part A of Title I required EPA to develop performance standards for new stationary sources. Stationary sources are defined by EPA as any building, structure, facility or installation which emits or may emit any air pollutant. Stationary sources are considered 'new' if constructed or modified after EPA issued the regulations for new performance standards. The New Performance Standards (NPS) are technology-based standards that apply to specific industrial processes and categories of stationary sources of air emissions such as solid waste landfills, steam generation, waste incinerators, cement plants, chemical plants and petroleum refineries, to name a few. The NPS regulations are found in 40 CFR Part 60 and the performance standards for specific facilities are located in the appendices to Part 60.

Another primary element of Part A of Title I is the NESHAPs. NESHAPs are regulatory standards for the emission of hazardous air pollutants (HAPs) from stationary sources that are known or suspected of causing cancer, other serious health effects or harmful environmental impacts. 40 CFR 61.01 contains a list of substances that have been designated as hazardous air pollutants and a list of additional substances that EPA has identified as potentially cancer-causing or having other serious health effects. NESHAPs establishes technology-based standards for stationary sources emitting HAPs. NESHAPs regulates two types of stationary sources: major sources and area sources. Major sources are defined as those stationary sources that either emit or have the potential to emit ≥ 10 tons per year of a HAP or ≥ 25 tons of any combination of HAPs. Area sources are defined as stationary sources that do not meet the definition of a major source. An important difference between the regulation of a major source and an area source is that major sources are required to meet EPA requirements for maximum degree of reduction in emissions of HAPs. These standards are known as *Maximum Achievable Control Technology Standards* (MACT). MACT standards are emission standards that are based on categories of industry that are already successfully controlling emissions of HAPs to low levels. EPA is required to evaluate MACT standards every eight years to determine if MACT standards should be revised to consider improvements in air control technology. Stationary sources subject to NESHAPs must initially test the performance air pollution control equipment to demonstrate compliance with the applicable emission limits. The initial performance test of air pollution control equipment is used to establish operating parameters of the equipment and its ability to control emissions of HAPs. Once operating parameters are established for a source, periodic monitoring is usually required to demonstrate ongoing compliance (U.S. EPA 2017). The EPA regulations for NESHAPs are found in 40 CFR Parts 61 and 63.

Part B of Title I was established to address ozone protection and was replaced by Title IV, *Stratospheric Ozone Protection*, as part of the 1990 amendments to the CAA.

The major components of Part C of Title I cover the prevention of significant deterioration of air quality. The purpose of the legislation under this part was to provide for the protection of air quality within air quality regions and to ensure that SIPs contained requirements for limiting emissions and other measures deemed necessary to prevent the significant deterioration of the air quality in areas meeting the NAAQS and designated as *attainment areas*. Part C also set forth permitting requirements for facilities being constructed after August 7, 1977, that would involve activities with emissions. Part C requires that these facilities could not be constructed prior to a permit being issued setting forth limits on emissions.

Part D of Title I is focused on nonattainment areas and the provisions for plans for achieving attainment of the national primary air quality standards. Each area proposed for designation as a nonattainment area must be published in the Federal Register to allow for public comment. Once an area is designated, the EPA is required to set a date that air quality standards must be obtained. Part D requires that this be as expeditious as possible but cannot exceed five years from the date the area was designated as a nonattainment area. Another important component of Part D was the authorization for states to set emission standards for motor vehicles, but only if they adopted standards that were equivalent to those established in the state of California (U.S. Congress 1977). This was unique in the fact that the CAA had limited the authority to regulate automobile emissions to the EPA and not to the states. California had developed progressive emission laws to address the serious air pollution problems within the state and was granted a waiver by Congress to regulate automobile emissions.

3.3.4.2 Title II

Title II of the CAA is divided into three parts:

- Part A – Motor Vehicle Emissions and Fuel Standards
- Part B – Aircraft Emission Standards
- Part C – Clean Fuel Vehicles

Title II authorized EPA to promulgate regulations focused on the regulation of air emissions from mobile sources. Part A of Title II authorized EPA to set standards for emission of pollutants from motor vehicles. Current regulations include standards for both new and older motor vehicles used on the highways and for off-road applications. This includes motor vehicles and engines powered by gasoline, diesel and all other fuels and additives used in motor vehicles. The majority of regulations relative to emissions from vehicles are found in 40 CFR Parts 85 and 86 and in 40 CFR Parts 89–94 for locomotive engines, off-highway internal combustions engines and marine vessel engines. Similarly, the regulations for fuels and fuel additives fall on the fuel refiners and those that are involved in the importation and distribution of fuels. The regulations for fuels and fuel additives are found in 40 CFR Parts 79 and 80 and include requirements for fuel quality, sampling, testing, blending, fuel components and labeling. The testing, use and distribution of fuels in nonattainment areas for ozone can be more rigorous due to the use of reformulated gasoline that is used seasonally to address periods of high ozone (U.S. EPA 2017).

Part B of Title II authorized EPA to establish emission standards for aircraft engines. The regulations set forth to control air pollution from aircraft are found in 40 CFR Part 87 and set emission standards, testing requirements and provisions for certification by aircraft engine manufacturers.

Part C of Title II established the emission standards for vehicles to qualify as clean fuel vehicles. These requirements identify performance standards for clean vehicle emissions of carbon monoxide, nitrogen oxides (NO_x), formaldehyde and nonmethane organic gas at 50,000 miles and 100,000 miles. There are specific emission standards for various vehicles based on gross vehicle weight and based on conventional, flexible and dual fuel uses. The clean fuel vehicle regulations are found in 40 CFR Part 88.

3.3.4.3 Title III

Title III of the CAA is titled *General Provisions* and includes legislation regarding federal procurements, emergency powers for the EPA and other administrative-related legislation that did not result in the promulgation of regulatory requirement applicable to facility, owners/operators, manufacturers, general public, etc. The most notable element coming from Title III was the requirement for the EPA to develop regulations for a national air-monitoring system using a standard air quality index. Today, the standard air quality index is used across the country by federal, state and local agencies that perform air monitoring and report air quality to the public. The standard air quality index is also used by agencies to control and/or limit emissions such as wood burning (no-burn days) during periods of low air quality.

3.3.4.4 Title IV

Title IV contains two elements: the first being noise abatement and the second being acid deposition control. The CAA directed the EPA to establish the Office of Noise Abatement and Control (ONAC) to investigate and research the effects of noise on public health. ONAC was in existence until 1981 when it was decided that noise abatement and control were best handled at the state and local levels (U.S. EPA 2019). As part of the 1990 CAA amendments, a new Title IV was added for acid rain but the original Title IV for noise pollution was not repealed even though ONAC no longer existed. Currently, the title for acid deposition control is Title IV-A. These amendments to the CAA established the acid rain program (ARP), which mandated significant reductions in sulfur dioxide and nitrogen oxides (NO_x) as the primary contributors to acid rain. The ARP has set a cap on the total amount of SO_2 emissions in the contiguous United States. The ARP final phase in 2010 established a national cap for the power sector of 8.95 million tons. The reduction of NO_x under the ARP has been focused on coal-fired power plants and utilizes more of a standard emission rate regulatory system. The emission cap instituted under the ARP led to the cap and trade program that is in place today for emission allowances. As part of the cap and trade system, emission sources are allotted allowances for emitting SO_2 and NO_x. Companies can buy and sell allowances as needed throughout the year. If an emission source has unused allowances, they can be retained for future use or sold on the allowance market. The system incentivizes emission sources to reduce SO_2 and NO_x emissions by improving emission controls and/or implementing other operational efficiencies so they do not have to use all of their allowances or buy additional allowances. The regulations for the ARP and associated processes are found in 40 CFR Parts 72–78.

3.3.4.5 Title V

Title V of the CAA is the section that covers the permitting program requirements for EPA and states. The 1990 CAA amendments revised the CAA permitting process so that all requirements applicable to an emission source (i.e., NESHAPs, NAAQS and NSPS) would be integrated into a single permit. The goal of this aspect of the amendment was to provide efficiencies in allowing states to implement and enforce CAA regulations. Similar to provisions under the CWA for NDES, authority can be granted to states to implement a permitting program once they have an EPA-approved program. EPA maintains jurisdiction for states, tribal areas and territories that do not have approved plans. An important aspect of the permitting program is that it allows for the collection of permitting fees to help offset the costs of program implementation. EPA's regulations for air-permitting programs are found in 40 CFR Parts 70 and 71 (U.S. EPA 2016).

3.3.4.6 Title VI

Title VI was added to the CAA as part of the 1990 amendments to develop programs for the protection of the stratospheric ozone layer. Programs for the protection of the ozone layer are focused on regulating and phasing out certain ozone-depleting substances (ODS). The phaseout of ODS was a direct result of the United States being party to an international treaty known as the Montreal Protocol. The Montreal Protocol focuses on the protection of the stratospheric ozone layer and calls for the phaseout of ODS. The U.S. EPA's ozone program took a multiphased approach to addressing ozone protection. The first phase was the phaseout of the most severe ODS compounds known as Class I ODS, which included chlorofluorocarbons, halons, methyl chloroform, carbon tetrachloride and methyl bromide. The phaseout schedule for Class I ODSs included prohibiting production and imports with a complete phaseout of all Class I ODSs by 1996 with the exception of methyl bromide, which was scheduled to be phased out in 2005. Phase II of the ozone protection program was the phaseout of Class II ODS or hydrofluorocarbons (HCFCs), which have a lower potential to affect the ozone as compared to Class I ODSs. The schedule for Class II ODS calls for a complete phaseout of HCFCs by 2030. There are certain regulatory exceptions to the phaseout schedule, which include ODS used in laboratory testing, certain agriculture-related uses and ODS that are reclaimed and specific HCFCs for certain uses (U.S. EPA 2018).

3.3.5 RESOURCE CONSERVATION AND RECOVERY ACT

RCRA comprises a set of regulations pertaining to the management and disposal of solid and hazardous wastes. One of the first set of laws to deal with wastes was the Solid Waste Disposal Act of 1965. This act was passed by Congress to address concerns associated with managing the increasing amounts of household and municipal wastes being generated. Similar to the early regulations for air pollution, the SWDA was focused on researching and understanding the problems associated with solid waste but had little success in addressing the growing waste problems (Eugene H. Robinson 2019). In 1976, Congress amended the SWDA with the passage of RCRA to address the concerns with waste management and disposal. One of the most important changes to RCRA occurred almost a decade later when Congress passed the Hazardous and Solid Waste Amendments (HSWA) in 1984, which made it illegal to land dispose of hazardous waste that did not meet the required treatment standards developed by EPA. RCRA was also amended in 1992 to increase RCRA regulatory authority at federal facilities under what is known as the Federal Facilities Compliance Act (FFCA) and again in 1996 with the Land Disposal Flexibility Act, which exempted certain wastes from the land disposal restrictions such as those being treated under a CWA permit (US Congress 1996).

3.3.5.1 RCRA Today

The RCRA regulation today is implemented in a similar fashion to the CAA and the CWA in that there is a permitting program and EPA has delegated implementation and enforcement authority to the states that have approved programs. RCRA regulates the management and disposal of both nonhazardous and hazardous wastes. Nonhazardous waste is commonly referred to as solid waste and includes the management and disposal of municipal wastes. The solid waste regulations are found in 40 CFR Parts 239–258 and are focused on the permitting and management of disposal facilities that accept for disposal nonhazardous waste from commercial, industrial and construction activities. Solid waste disposal facilities must be located in areas that are conducive to waste disposal and the long-term protection of the environment. Solid waste disposal facilities are required to have approved designs for how the landfill is constructed to include impermeable barriers that provide for containment of leachate and monitoring systems to detect releases of contaminants that could potentially impact surface or groundwater. Solid waste disposal facilities are required to obtain permits that define the operating parameters for the landfill. Permits include requirements regarding the types of waste that can be accepted for disposal, criteria for accepting incoming waste, requirements for covering the waste, monitoring requirements, record-keeping, regulatory reporting and requirements for closing the landfill when it is no longer operational. The solid waste regulations under RCRA are most relevant to solid waste landfill owners and operators. Companies that send their nonhazardous wastes to a solid waste landfill need to understand the specific types of waste the landfill can receive and the processes for getting approval to send wastes to the landfill.

The hazardous waste regulations under RCRA apply not only to facilities that receive hazardous waste for disposal but also to those that generate, package, store, treat and transport hazardous wastes. RCRA has created what is known as a 'cradle-to-grave' management system for hazardous waste, where required documentation ensures that waste is tracked from the point of generation to the final disposal, and the waste generator is ultimately responsible for their waste until it is disposed. RCRA also includes regulations pertaining to USTs and used oil.

3.3.5.2 Waste Generators

Generators of regulated hazardous wastes are required to characterize the physical and chemical properties of their waste to determine the applicable requirements for managing the waste. To be classified as hazardous, a waste must be a solid waste which basically includes any discarded material that is being abandoned, disposed of, burned, recycled or accumulated in lieu of being disposed. There are some exceptions to the definition of a solid waste, including domestic sewage, wastewaters that are point sources and discharged under the CWA and radioactive material regulated under the Atomic Energy Act (AEA). Solid wastes are hazardous wastes if they exhibit a hazardous characteristic (characteristic waste) as defined by EPA or contain a constituent that is specifically identified by EPA as hazardous (listed waste). Characteristic wastes are those that are ignitable, corrosive, reactive or toxic. These wastes are defined in 40 CFR Parts 261.20–261.24. Listed wastes are wastes produced from specific and nonspecific sources or wastes containing specific chemicals that are commercial chemical products, off-specification chemicals being discarded or any residues or spills of these materials. Listed wastes are described in 40 CFR Parts 261.30–261.33. One critical consideration for companies that generate hazardous waste is that RCRA regulations stipulate that any nonhazardous waste mixed with a hazardous waste by definition becomes a hazardous waste. Known as the 'mixture rule,' this aspect of the regulations can result in significant costs to a company that fails to segregate hazardous and nonhazardous wastes whenever possible as the costs for managing hazardous waste far exceed those for nonhazardous waste. Generators are also required to determine and monitor the volume of waste they generate on a monthly basis as the regulations vary based on the volume of waste a facility generates.

A facility that generates 100 kilograms or less of hazardous waste and 1 kilogram or less of acutely hazardous waste per month is considered a very small quantity generator (VSQG). VSQGs must identify all hazardous waste they generate and must send the waste to a facility that is authorized to handle the waste. VSQGs may not accumulate more than 1,000 kilograms onsite at any one time.

A facility that generates more than 100 kilograms but less than 1,000 kilograms per month is considered a small quantity generator (SQG). SQGs can accumulate up to 6,000 kilograms of waste onsite for 180 days without a permit. This timeframe is extended to 270 days if the facility has to ship the waste more than 200 miles. SQGs must meet the requirements in 40 CFR 262.16 to include the container labeling requirements, manifesting hazardous waste shipments and the land disposal restrictions.

A facility that generates ≥1,000 kilograms of hazardous or >1 kilogram of acutely hazardous waste per month is considered a large quantity generator (LQG). LQGs can accumulate waste onsite for only 90 days without a permit and are subject to a larger set of requirements than VSQGs and SQGs. The requirements for LQGs are found in 40 CFR 261.17.

3.3.5.3 Hazardous Waste Transporters

Transporters that haul hazardous waste must obtain a unique EPA identification number (EPA ID) from EPA prior to transporting hazardous waste. Transporters are required to adhere to all waste manifesting requirements and cannot accept any hazardous waste from a generator without a hazardous waste manifest. Transporters must adhere to the transfer facility, manifesting record-keeping and immediate actions requirements in 40 CFR Part 263.

3.3.5.4 Treatment, Storage and Disposal Facilities (TSDFs)

Facilities that store, treat and/or dispose of hazardous waste are required to meet a host of requirements depending on the nature of the hazardous waste activities occurring at a facility. The following subparts of 40 CFR Part 264 identify the requirements that apply to TSDFs:

- Subpart B – General Facility Standards: These standards apply to all facilities that treat, store or dispose of waste that aren't otherwise exempted (i.e., VSQG or SMQ storing waste less than 90 days).
- Subpart C – Preparedness and Prevention: This subpart requires that TSDFs are designed, operated and maintained to minimize the possibility of fire, explosions or unplanned releases.
- Subpart D – Contingency Plan and Emergency Procedures: This subpart requires TSDFs to prepare and maintain contingency plans and emergency procedures which provide details for responding to emergencies, including emergency equipment, points of contact and agreements with local emergency response organizations such as police, hospitals and fire departments.
- Subpart E – Manifest System, Record-Keeping and Reporting: This subpart identifies the requirements for waste manifesting, the facility operating record, records availability and retention and regulatory reporting.
- Subpart F – Releases from Solid Waste Management Units: In general, these regulations apply to landfills, surface impoundments, waste piles or land treatment units, and they identify the requirements for detecting, characterizing and responding to releases into an aquifer.
- Subpart G – Closure and Postclosure: This subpart contains the requirements for closing a TSDF, the associated performance standards and the required postclosure actions.
- Subpart H – Financial Requirements: This subpart requires owners/operators of TSDFs to maintain estimates of the cost to close the facility and postclosure maintenance and to establish financial assurance for closure of the facility.
- Subpart I – Use and Management of Containers: This subpart identifies the requirements for containers at all hazardous waste facilities that use containers to store hazardous waste.

- Subpart J – Tank Systems: This subpart identifies the requirements for tank systems at all hazardous waste facilities that use tanks to store hazardous waste.
- Subparts K through X identify the requirements facilities engaged in specific methods of waste management such as surface impoundments, waste piles, land treatment, landfills, incinerators, Corrective Action Management Units (CAMUs), Drip Pads and miscellaneous units.
- Subparts AA through CC identify the requirements for controlling air emissions from process vents, equipment leaks and emission standards for tanks, surface impoundments and containers.
- Subpart DD – Containment Buildings: This subpart identifies the design and operating standards for completely enclosed containments buildings.
- Subpart EE – Hazardous Waste Munitions and Explosive Storage: This subpart contains the requirements regarding the design, operation and closure of the storage units or magazines.

This discussion presents a simple regulatory summary of the requirements for TSDFs storing, treating or disposing of hazardous wastes. The RCRA regulations for TSDFs are extensive and complex. With only a few exceptions, most TSDF owners/operators will be required to obtain a permit that will have an integrated set of applicable requirements that define the operating parameters and compliance obligations for the TSDF.

3.3.5.5 Land Disposal Restrictions (LDRs)

As discussed earlier in this section, the 1984 HSWA made it illegal to dispose of hazardous waste without it being treated to meet the treatment standards to be developed by EPA. LDRs identify the treatment standards that hazardous waste must meet before it can be placed on land (i.e., disposed in a landfill, surface impoundments, waste pile). LDRs apply to generators of hazardous waste and to facilities that store, treat or dispose of waste. There are a few exceptions to the applicability of the LDRs, including VSQGs and wastes that are treated under the CWA requirements. 40 CFR 268.40 contains the treatment standards for hazardous waste presented in tabular form, which identifies the waste code, associated hazardous constituents and the applicable treatment standard for that constituent waste type. The treatment standards table will identify the treatment standard for wastewaters and nonwastewaters and either a concentration-based or a technology-based standard. Concentration-based standards identify the concentration of a specific constituent that must be achieved before the waste can be land disposed. The concentration-based standards require that the waste either be analyzed directly or that an extract of the waste be analyzed depending on the waste constituents. Other waste constituents are required to be treated by a specified technology such as combustion or chemical oxidation, just to name a few examples. It is important to note here that all waste analyses must be conducted using specific analytical methods. Debris that has been contaminated with hazardous constituents (wood, concrete, etc.) present a unique set of challenges related to meeting LDRs. Alternative treatment standards for hazardous debris are identified in 40 CFR 268.45. The treatment standards for debris are based on treatment technologies such as physical extraction (e.g., grinding, high-pressure washing), chemical extraction, thermal extraction, chemical destruction (oxidation, reduction and thermal) and immobilization technologies including micro- and macroencapsulation. For identified technology, the debris treatment standards specify the performance standard and any contaminant restrictions that may apply (U.S. EPA 2019).

There are several important considerations for owners/operators of facilities that generate hazardous waste subject to LDRs: (1) characteristic wastes that are treated and no longer exhibit the characteristic that made it hazardous can be disposed of in a nonhazardous landfill, (2) waste cannot be diluted as a substitute for meeting the LDR treatment standards and, (3) waste not meeting LDR cannot be stored onsite as a means to avoid treating the waste. In general, waste cannot be stored for more than one year without approval from the appropriate regulatory agency.

3.3.5.6 Universal Waste

Universal wastes are batteries (excludes lead-acid batteries), recalled or canceled pesticides, lamps (fluorescent, neon, mercury vapor sodium, etc.) and mercury-containing equipment. These wastes are considered universal as they are generated by many different companies and businesses. EPA developed a streamlined set of requirements for universal wastes to promote recycling of these wastes and to encourage the reduction of these types of wastes going to landfills. These streamlined regulations include the ability to store universal wastes for a year, they do not require a manifest for shipping and reduced labeling requirements and universal waste volumes do not have to be accounted for in determining generator status (i.e., VSQG, SQG, LQG). Universal wastes have to be managed to prevent releases and sent to a facility that is approved or permitted to receive these wastes types. States can adopt the universal waste regulations as is, choose to regulate fewer wastes as universal or add additional universal wastes. The regulations for universal waste are found in 40 CFR Part 273 and include requirements for universal waste handlers, transporters and universal waste destination facilities (U.S. EPA 2019).

3.3.5.7 Underground Storage Tanks

Under the RCRA regulations, USTs are defined as tank and any associated piping with 10% or more of its volume underground and that is used to hold petroleum or hazardous substances (includes hazardous wastes). Prior to the enactment of the UST regulations in the 1980s, there were no requirements for owners/operators of facilities with USTs to monitor for leaks, verify the integrity of the tanks or piping or take any actions to prevent deterioration of the UST system and mitigate future releases. Up until the passage of the UST regulations, there were literally millions of USTs in use and a large percentage were leaking contaminants into the soils and groundwater. To address these concerns, Congress passed the SWDA in 1984, which required EPA to develop UST regulations with standards for tank design, installation, leak detection, spill and overflow control, corrective actions and tank closures. In 1986, the SWDA was amended under the Superfund Amendments Reauthorization Act, which resulted in the establishment of financial responsibility requirements for UST owners/operators to cover cleanup and third-party liability costs. These regulations are in 40 CFR Part 280 and identify the technical requirements for tank design and installation, corrosion protection, spill/overfill prevention, leak detection, tank tightness testing and responding to and cleaning up releases. The UST regulations also include requirements for UST owners/operators to have the financial means to cover costs of a cleanup, a release and compensate for damages to third parties. The financial responsibilities can be met by a variety of mechanisms, including insurance coverage, corporate bonds/guarantees or financial assistance funds provided by some states. The UST regulations also provide requirements that owners/operators must adhere to when closing or taking a tank out of service. As with the hazardous waste program authority, EPA has delegated the authority to implement and enforce the UST program to those states with approved programs (U.S. EPA 2017).

3.3.5.8 Used Oil

EPA's regulations for used oil are based on the stated presumption that used oils (defined as oil derived from crude or synthetic oils that have been used) will be recycled unless they are sent for disposal. Used oil can be reconditioned by removing unwanted constituents, it can be re-refined by reintroduction into the petroleum refinery processes or it can be burned for energy recovery in certain industrial applications as allowed by the regulations. The used oil regulations set forth the requirements for storage, labeling, spill response and record-keeping. Owners/operators of facilities that ship used oil must use a transporter that has a valid EPA ID number. Also, transporters and facilities receiving used oil for processing or transfer are required to keep a record of each time they accept use oil into their facility or for transport. An important consideration in managing used oil is that it can become subject to hazardous waste regulations if it is mixed with a hazardous waste. Once used oil is mixed with a hazardous waste, it becomes subject to the hazardous waste requirements pertaining to storage, treatment and disposal and the significantly higher associated costs (U.S. EPA 2019).

3.3.6 EMERGENCY PLANNING AND COMMUNITY RIGHT TO KNOW ACT (EPCRA)

EPCRA was enacted in 1986 as part of the Superfund Amendments Reauthorization Act (SARA) and in part was in response to a toxic chemical release in Bhopal, India, that killed more than 3,000 people immediately and caused serious injuries and lingering health effects to thousands of others (Broughton 2005). In an effort to preclude this type of disaster from occurring in the United States, Congress passed EPCRA as a mechanism for emergency planning and informing the public of chemicals being stored at facilities within their communities. EPCRA requires involvement at the state and local levels in order to protect citizens from chemical hazards that are stored within their communities. Under EPCRA, states are required to establish a State Emergency Response Commission (SERC), which is responsible for establishing emergency planning districts within their state. Each emergency planning district within a state has a local emergency planning committee (LEPC) that includes members from organizations such as emergency responders, local government officials, community groups and industry representatives (U.S. EPA 2018). EPCRA is set up to require facilities storing chemicals to report the types of chemicals and quantities to state and local officials who in turn use the information to prepare chemical emergency response plans.

Facilities have several potential reporting requirements under EPCRA. First, under the 'community right-to-know' requirements, facilities that handle or store chemicals are required to submit an inventory listing of their chemicals, locations and the associated safety data sheets (SDSs) to state and local officials as well as the local fire department. This information is submitted using a 'Tier I' or 'Tier II' inventory report. A Tier I form contains general information about the chemical stored at a facility, whereas a Tier II form has more specific information about chemical inventories and locations. Most states require facilities to use the Tier II form for reporting.

Facilities may also need to submit an annual Toxic Release Inventory (TRI) report if they process, use or store certain toxic chemicals that can cause cancer, acute or chronic human health effects or significantly impact the environment. There are currently 595 chemicals covered under the TRI program, each with established reporting threshold levels. The TRI report identifies how much of each applicable chemical is released into the environment, recycled or land disposed. Owners/operators of facilities that must submit a TRI report are identified in the regulation's North American Industry Classification System (NAICS) codes. These include industries such as hazardous waste treatment, manufacturing, mining, power generation and federal facilities (U.S. EPA 2019).

Facilities that handle or store chemicals classified as extremely hazardous substances in quantities greater than established thresholds [threshold planning quantities (TPQs)] must cooperate with state and local agencies in emergency planning and are required to report any accidental release above established reportable quantities (RQs) immediately to state and local agencies (U.S. EPA 2018).

3.3.7 TOXIC SUBSTANCES CONTROL ACT

TSCA was another environmental law that was enacted by Congress during the environmentalism movement in the 1970s. Enacted in 1976, TSCA required the EPA to develop regulations pertaining to the manufacturing, importation, use and disposal of specific chemical substances. The TSCA regulations found in 40 CFR Parts 723–770 contain the testing, reporting and record-keeping and other requirements for chemical substances. A primary intent of TSCA was to regulate the manufacture, distribution, use and disposal of toxic chemical substances or mixtures in order to mitigate unreasonable risks. This includes the requirement for manufactures to notify the appropriate regulatory agency regarding the development of new chemical substances or mixtures.

TSCA also regulates a few specific chemicals in a somewhat different manner. One of the most commonly known chemicals regulated under TSCA is the polychlorinated biphenyls (PCBs). EPA has established requirements in 40 CFR Part 761 that regulate the manufacture, distribution, storage, cleanup, transportation, reporting and disposal of PCBs. PCBs have been around since the late 1920s and have been used in a variety of items, including hydraulic fluids, paints, dielectric fluids and sealants, to name a few. The production of PCBs was banned in 1979 due to the associated toxicity of PCBs and their persistence in the environment. Even though the production of PCBs has remained banned for decades, PCBs are still commonly found in the environment because they have been used in countless industrial and commercial products and are so persistent in the environment.

TSCA includes regulations in 40 CFR Part 763 pertaining to asbestos in schools, which require educational agencies to identify asbestos-containing material in schools and to develop and implement asbestos management plans in coordination with the state governor and the state and local agencies. Another common element of TSCA is the lead-based paint regulations found in 40 CFR Part 745. These regulations were developed to address the presence of lead-based paint in houses constructed before 1978 and where children may be present. The lead-based paint regulations include requirements for the inspection, training, identification, notification, disclosure and work practices for repairs and renovations. The lead-based paint regulations developed to ensure that the health hazards associated with the presence of lead-based paint were identified, occupants were properly notified and appropriate remedies could be taken to protect children and others who could potentially be exposed (U.S. EPA 2019).

3.3.8 Safe Drinking Water Act (SDWA)

The SDWA was originally passed by Congress in 1974 for the purpose of establishing regulations for the protection of public drinking water. The SDWA established national drinking water standards that set maximum contaminant levels (MCLs) for substances identified by EPA as potentially harmful to human health and that may be present in drinking water. Public drinking water systems are those that serve at least 25 people for 60 days a year or more. EPA reports that there are more than 170,000 public water systems providing drinking water to the public (U.S. EPA 2004). The SDWA was amended in 1986 to finalize the national primary drinking water standards that were established as interim standards after the original SWDA was enacted in 1974. The SDWA underwent a major amendment in 1996 to expand the requirements of the act to include protections for sources of drinking water, certification requirements for operators of public drinking water systems and requirements for public drinking water systems to provide information to consumers annually about where the water comes from, the quality of the water provided, information about water testing, any contaminants that have been detected and health-related information associated with any detected contaminants. Programs were also established where states could receive funding through federal grants to assist with water system upgrades and protection activities for source waters.

All but two states have jurisdictional authority to implement and enforce the SDWA requirements. Wyoming, Washington, DC, and tribal lands fall under the jurisdiction of the U.S. EPA. States with approved programs are required to develop, implement and enforce drinking water regulatory programs that are at least as stringent as the federal regulations. States that are authorized under the SDWA must establish programs to protect water sources that supply public drinking water systems. Water sources can include surface waters (lakes, rivers, reservoirs) and groundwater. Many of these protections for surface and groundwater are accomplished through other environmental regulations such as the CWA. The SDWA provides additional protections through a program known as the Underground Injection Control (UIC) program that regulates the injection of wastes that could impact groundwater.

The EPA and the states are responsible for overseeing public drinking water utility companies to ensure SDWA requirements are met, including the following:

- System operators are certified
- Water quality standards are being met
- Approved treatment and disinfection systems are in place as necessary to ensure MCLs are not exceeded
- Water testing and analysis requirements are met
- Annual consumer confidence reports are submitted
- Water system and associated infrastructure is maintained
- Assessment and protection of source water

Public drinking water systems that violate drinking water regulations are subject to enforcement by the EPA or state agencies under the authority provided by the SDWA (U.S. EPA 2017).

3.3.9 NATIONAL ENVIRONMENTAL POLICY ACT

NEPA was originally passed by Congress in 1969 with the intent of establishing a national policy that promoted a productive and mutually congruent relationship between man and the environment and emphasized a needed balance between the forward progression of human society and protection of the environment. The act also established the CEQ, which was to advise the president and work with federal agencies on matters regarding protection of the environment. To further promote this national policy, NEPA set forth requirements that each federal department or agency must, in consultation with the CEQ, provide a detailed report that considers the environmental impacts of proposed major federal actions, the unavoidable impacts resulting from implementation and the alternatives to the proposed action. NEPA was amended in 1970 under the Environmental Quality Improvement Act to establish the Office of Environmental Quality as an office under the Executive Office of the President. The Office of Environmental Quality would later be dissolved but set the framework for the establishment of the EPA (CEQ 2019). However, the CEQ still exists today and is responsible for ensuring that federal agencies are meeting the requirements set forth in NEPA.

NEPA requires that federal agencies and departments develop procedures for evaluating the environmental, social and economic impacts of proposed major actions. Federal actions subject to NEPA are those that are financed (partially or entirely), assisted, conducted, regulated or approved by a federal agency. These actions include the adoption of regulations/policies, formal plans, programs to implement plans/policies and approval of specific projects located in the geographical area. As part of the evaluation of impacts, federal agencies must prepare a detailed report that provides information regarding their assessment of the proposed activity, the associated environmental impacts and alternative actions. This information is required to be provided to the public for review and comment.

There are three levels of evaluation or analysis provided for under NEPA depending on the potential of the proposed project to have significant environmental impacts. The first level is where a federal action can be excluded from a detailed analysis, because the proposed action is determined to not have a significant impact on the environment. This level of analysis is called a categorical exclusion (CATEX), and the processes for determining which proposed actions qualify for a CATEX are proceduralized at the federal agency level and each determination is documented and maintained as a record.

The next level of analysis is called an Environmental Assessment (EA). An EA is used to determine if the proposed action has potential to cause significant environmental impact. An EA will include discussion relative to purpose and need of the proposed action, any environmental impacts associated with the action or proposed alternatives and a listing of other agencies or experts consulted during the assessment. EAs determine whether or not a proposed project will have significant environmental impacts. If the determination is that there will be no significant impacts, the federal agency documents the reasoning for the determination in a Finding of No Significant Impact (FONSI). If, on the other hand, the EA determines that there will be significant environmental impacts, then the agency must prepare an environmental impact statement (EIS).

An EIS is the most comprehensive level of analysis conducted under NEPA and involves a substantial level of public participation. Prior to preparing an EIS, the federal agency proposing the action must publish a Notice of Intent (NOI) in the Federal Register, informing the public of the pending environmental analysis and how they can participate. The NOI constitutes the commencement of the scoping process where the potential issues and alternatives that need to be addressed in the analysis are identified by federal agency with input from the public and other relevant federal, state and local agencies that have regulatory responsibilities or expertise associated with the proposed action. Upon completion of scoping, a draft EIS is prepared and made available for a public review and comment period of no less than 45 days. Agencies are required to evaluate and consider all substantive comments. If comments received by the public warrant further analysis by the agency, then additional analysis is conducted prior to finalizing the EIS. Preparation of the final EIS will incorporate responses to substantive comments and any additional analysis as necessary. Once the EIS is finalized, there is a 30-day waiting period that normally takes place before an agency makes a final decision on the proposed action. Once the EIS is finalized, a notice of availability of the final draft of the EIS is published in the Federal Register. The federal agency's final decision on the proposed action is documented in a Record of Decision (ROD) that provides an explanation of the decision, the alternatives that were considered and any other actions that may be necessary to alleviate and/or monitor impacts. If the scope of the proposed project changes or new information becomes available once the EIS is finalized, the federal agency may have to prepare a supplemental EIS to address any new potential impacts. A supplemental EIS basically follows the same process as the original EIS (U.S. EPA 2017).

3.3.10 ENDANGERED SPECIES ACT

The ESA was passed by Congress in 1973 and U.S. Fish and Wildlife with the purpose of protecting endangered species and their habitats and to promote population recovery and stabilization. The ESA is implemented by the U.S. Fish and Wildlife Service (FWS) and the National Marine Fisheries Service (NMFS) under the National Oceanic and Atmospheric Administration (NOAA). The FWS has responsibility for terrestrial and freshwater species and the NMFS has responsibilities for the protection of marine wildlife. While these two agencies have the lead roles in the ESA, all federal agencies are required to use their authority to promote the ESA and to consult with FWS or NMFS on any proposed actions to ensure that endangered species and/or critical habitat are not adversely affected. Authority to implement and enforce the requirements of the ESA is not delegated to the states, although federal funding is made available to encourage states to develop programs for protecting endangered and threatened species (U.S. FWS 2018).

EPA's role in protecting endangered species is accomplished under the Endangered Species Protection Program (ESPP). The ESPP evaluates the impacts of pesticides on endangered species and their habitats when reviewing new pesticides for registration. Pesticides that are determined to be harmful to protected species or critical habitat may have limitations placed on how and where they can be used as part of the EPA registration process. EPA's authorization for this role is derived from the FIFRA, which is discussed further below (U.S. EPA 2018).

Species are protected under the ESA if they are considered to be in danger of extinction (endangered) or they are likely to become endangered in the near future (threatened). FWS and NMFS evaluate several factors when considering a species for listing as an endangered or threatened species under the ESA, including damaged habitat; overuse of a species for commercial, recreational or other purposes; disease; predation; the need for additional protective measure and other relevant factors. Once a species is listed as endangered or threatened, any critical habitat important to the protection and recovery of the species is identified by FWS or NMFS. Habitat that is designated as critical requires actions by federal agencies to avoid 'destruction' or 'adverse modification' of designated critical habitat. Designation of a species for listing under the ESA requires the notice of designation to be published in the Federal Register and made available for public review and comment. Designation of critical habitat for the species must also follow this same process (NOAA 2019).

Once a species is designated as endangered or threatened, a recovery plan is developed, which describes the steps for recovering the species and stabilizing populations. The ESA prohibits the 'take' of any species protected under the law. A take is defined as to harass, harm, pursue, hunt, shoot, kill, trap, capture or collect. The word harm in the definition means to kill or injure and also includes modifying or degrading habitat such that it injures or kills the protected species. Any activities conducted by a federal agency that may result in the 'take' of a protected species must be approved by FWS or NMFS, which will identify the specific controls and conditions regarding the incidental 'take' of a protected species. The approval for incidental takes are referred to as 'take permits' and are commonly documented in a biological opinion issued by one of the services. Nonfederal entities and private landowners wishing to develop or modify property inhabited by protected species can develop a Habitat Conservation Plan (HCP) in consultation with FWS that, once approved, will allow for the incidental take of a species resulting from the development activities. HCPs include a description of the activity, an evaluation of the potential impacts to the species, steps that will be taken to minimize or avoid impacts to the species and demonstration of adequate funding to carry out the actions (U.S. FWS 2018).

Under the ESA, conservation land banks are utilized to offset the unavoidable loss of critical habitat. This concept is similar to the conservation banks utilized under the CWA for wetlands mitigation. Conservation banks are lands that are permanently protected for one or more protected species and are used as conservation credits for loss of habitat and/or protected species at some other location. Similar to the CWA mitigation banks, agencies or landowners that have adversely impacted critical habitat or protected species can purchase mitigation credits from a conservation bank to mitigate for the loss of habitat or species. Conservation credits are purchased from a land bank that is permanently protecting the same species that was impacted by the activity requiring mitigation (U.S. FWS 2013).

3.3.11 Federal Insecticide, Fungicide and Rodenticide Act

As discussed earlier in the chapter, FIFRA was enacted in 1947 to address concerns regarding the unintended effects of pesticides on human health, nontargeted species and the environment by establishing labeling requirements for pesticides that were sold and distributed commercially. FIFRA was originally the responsibility of the U.S. Department of Agriculture until it was transferred to the newly formed EPA in 1970. In1972, FIFRA was amended by the Federal Environmental Pesticide Control Act to authorize regulation of the use of pesticides and stronger enforcement. The objective of FIFRA regulations is to ensure that pesticides are evaluated prior to being distributed and commercially available and that their use will not cause risk to human health, protected species or the environment. The EPA regulations for pesticides as authorized under FIFRA are found in 40 CFR Parts 150–180. A large portion of the FIFRA regulations pertains to those that manufacture, package, transport, distribute or sell pesticides. As discussed under the ESA section above, the regulations require that all pesticides be registered with the EPA before they are distributed or sold. Pesticides that pose an unreasonable risk to human health or the environment are prohibited from registration by the EPA. Once a pesticide is registered with EPA, any use restrictions required by EPA must be adhered to. Other regulatory aspects of FIFRA include the following:

- Pesticides intended for residential use must be packaged in child-resistant packaging.
- All pesticides must be labeled in accordance with 40 CFR Part 156.
- Facilities that use or handle pesticides must have established work processes that minimize or preclude exposure and must have emergency response procedures regarding exposures.
- Pesticides designated by EPA as 'restricted use' can be used only by a certified applicator or someone under their supervision. Certification is provided by EPA or by states that have an EPA-approved certification program for pesticide applicators.

States have the option of regulating registered pesticide use or sales within the state, but they cannot require any additional labeling or labeling that is different than FIFRA requirements nor can they authorize the sale or any use that is prohibited under FIFRA (U.S. EPA 2019).

3.3.12 COMPREHENSIVE ENVIRONMENTAL RESPONSE COMPENSATION AND LIABILITY ACT (CERCLA)

CERCLA, commonly referred to as Superfund, was passed by Congress in 1980 in response to the presence of a large number of locations that were contaminated with toxic wastes that were being abandoned by those responsible. CERCLA provided authority to the EPA to require responsible parties to clean up abandoned contaminated sites and active sites with uncontrolled releases of hazardous substances. For contaminated sites that pose immediate threats to human health and the environment or a responsible party cannot be identified, the EPA can cleanup using site funds from a trust fund established under CERCLA. Funded by a tax on chemical and petroleum industries, the trust fund collected more than 1.5 billion dollars over a five-year period and provided a large amount of funding to be used for the immediate and long-term cleanup of contaminated sites. EPA was also authorized under CERCLA to pursue cost recovery actions for actions taken by the agency to clean up contaminated sites that were either abandoned or active sites with uncontrolled releases of hazardous waste. Contaminated sites that require cleanup but do not pose an immediate threat to human health are identified by EPA and placed on the National Priorities List (NPL). The NPL provides a listing of contaminated sites requiring long-term remedial actions to clean up existing contamination to prevent ongoing releases of hazardous substances and mitigate impacts to human health and the environment. Sites that are subject to a remedial action under CERCLA will be working under the direction of the EPA and/or other regulatory agencies when required to conduct immediate removal actions or long-term site remediation. These actions are typically done under a consent order and for NPL sites involve site investigations, feasibility studies, remedial action plans and a record of decision, documenting the required corrective actions and long-term monitoring (U.S. EPA 2018).

3.4 STATE LAWS

Environmental regulations at the state levels are most often driven by the federal regulations authorized by Congress under a specific law or act. The major environmental laws passed by Congress require the EPA to develop regulations to address the aspects of the law. As an example, when Congress passed the CWA, EPA was required by the law to develop regulations for protecting waters of the United States. The CWA also provided EPA with the authority to implement and enforce the requirements of the CWA. Once enacted by Congress, federal laws apply to all states and territories of the United States. The majority of the environmental regulations developed by EPA contain provisions for states to obtain authorization to implement and enforce the regulations within their state. Federal laws such as the CWA, CAA, RCRA and SDWA all have provisions in the associated regulations developed by EPA for states to develop authorized programs or obtain primacy. States can develop programs for implementing a specific environmental regulatory program and pursue EPA approval of the program. Once a state's program is approved by EPA, they are authorized to implement and enforce the program within their state. EPA is the regulatory authority in states that have not sought authorization and/or have not obtained EPA approval. Most states have been authorized by EPA to implement the above-mentioned programs within their states, while the EPA maintains jurisdiction over tribal lands and a handful of states that do not have approved programs.

State program development includes legislative actions to incorporate the federal regulations into state statutes followed by the passage of state regulations for implementing and enforcing the regulations. States can either adopt the federal regulations by reference or add additional state requirements, but state regulations must be as stringent as the federal regulations at a minimum. Federal law will take precedent when there is a conflict between state and federal law. As an example, many states with authorized programs under RCRA for hazardous waste have chosen to implement state regulations that mirror the federal regulations, while other states have chosen to implement a more rigorous set of hazardous waste laws that include additional requirements not mandated by federal law. The decision for a state to implement additional requirements beyond what is required at the federal level can be based on state-specific issues or concerns as well as the types of industries operating within a state. As an example, air and water quality concerns within a specific state can constitute a reason for more rigorous regulations pertaining to air emissions and/or point source discharges.

Environmental regulations such as CERCLA, ESA and NEPA do not have the same type of mechanisms for delegation to the states. Both NEPA and ESA encourage states to develop programs, but state programs developed under these two regulatory programs are in addition to the federal requirements. The CERCLA (aka Superfund) does not contain provisions for EPA to delegate authority to the states, although EPA cooperatively works with state regulatory agencies regarding the remediation of contaminated sites.

Part of the strategy for a successful compliance program is understanding which regulations apply to your organization. In order to accomplish this successfully, it is critically important to understand whether your state is authorized by EPA to implement specific regulations, do the state regulations mirror the federal regulations or are there additional requirements, and which state agency is responsible for specific regulatory programs.

3.5 COUNTY LAWS

Many states will delegate the authority to implement specific regulatory programs to counties. To use the hazardous waste regulations again as an example, states that are authorized by EPA to implement and enforce the RCRA hazardous waste regulations may further delegate oversight authority to agencies at the county level. Delegation of authority to a county agency means that the designated agency will be responsible for performing inspections, oversight and enforcement of regulated facilities within their county. The oversight of municipal waste landfills is another good example of where a federal regulation that has been delegated to the states is commonly further delegated to the county level. Much like when EPA delegates program authority to the states, state delegation to a county agency does not preclude the state agency's right to pursue enforcement or legal actions against noncompliant facilities. While delegation from the states to the county level is common, it can widely vary from state to state based on factors such as population, available funding and size of the regulated community (i.e., number of hazardous waste generators, stationary air emission sources, wastewater discharges).

Laws and regulations may be enacted within a state that are not driven by federal laws or regulations but are necessary to address state-specific issues or concerns. These state-specific environmental laws may include implementation and/ or enforcement roles that are required to be administered by county agencies. The management of hydrocarbon-contaminated soils is one example where state-specific action levels and cleanup standards can be implemented and enforced at the county level. Septic systems are also commonly regulated at the county level where the design and installation of septic systems typically require permitting and oversight by a county agency.

3.6 LOCAL LAWS

Local laws are often a further delegation of authority from the states and counties. Locally owned POTWs are operated and managed by local municipalities, and these local municipalities are authorized and responsible for ensuring the pretreatment standards under the CWA are met by nonresidential wastewater dischargers. POTWs utilize a permitting system to manage what constituents and associated concentrations may be discharged to the sanitary sewer system. Permittees discharging to a POTW are required to meet discharge permit requirements which include ongoing periodic analysis of wastewaters to demonstrate compliance to the permit. Local municipalities are responsible for implementing the permitting process, providing oversight of permittees and have the regulatory enforcement authority to pursue fines and penalties or other legal actions against permittees who violate permit conditions. Because storm water runoff is often collected and managed by local municipally owned systems, municipalities may also impose local requirements for storm water management consistent with their MS4 permits. There are often recycling and waste reduction policies which are implemented at a local level where city government agencies impose regulations regarding the management and disposal of specific items such as plastic bags, Styrofoam food containers and used tires. Many of these local environmental regulations and policies are put in place as measures to control these types of problematic waste items because either they aren't regulated or existing higher-level regulations are inadequate.

3.7 REGULATORY RELATIONSHIP DEVELOPMENT AND SUSTAINMENT

Facility owners, operators and managers responsible for environmental compliance need to effectively identify applicable regulations, understand the appropriate regulatory agency for each applicable regulation and be knowledgeable of agency personnel with oversight responsibilities at their facility. With the availability and accessibility of the Internet today, regulatory information is readily available. The majority of federal, state, county and local agencies have websites that are specifically designed to make regulatory information available to the regulated communities. The EPA has a voluminous amount of regulatory information available on their website that provides information relative to the history, current status of environmental laws and regulations, proposed regulations, points of contact for the various regulatory programs and in many instances information on state programs.

State, county and local environmental regulatory agencies also have a great deal of regulatory information readily available on the Internet, providing not only regulatory information but also agency points of contact, organizational structures as well as guidance documents, checklist and other tools designed to assist facilities in assessing regulatory applicability and maintaining compliance. Many agencies offer or participate in classroom and online seminars or training that can provide invaluable insight to specific regulatory programs (i.e., air permitting, NPDES permits, hazardous waste) and provide opportunities to interact with regulatory agency personnel. While the Internet has greatly enhanced the ability of facility owners/operators and compliance personnel to readily access a vast amount of regulatory information, it does not replace the need for establishing a good working relationship with regulatory agency personnel providing oversight at your facility.

Based on the complexity of operations at a given facility, environmental compliance can range from meeting a single regulation involving little or no regulatory oversight to highly complex operations involving numerous regulations, oversight from multiple regulatory agencies and various permitting, reporting and regulatory notification requirements. Regardless of complexity, one of the most critical elements of a successful and sustainable compliance program is developing and maintaining good working relationships with the regulators. Establishing professional and constructive working relationships with regulatory personnel such as inspectors, permit writers and regulatory program managers is most often mutually beneficial to both parties. This type of beneficial relationship provides for open communication, mutual respect for the regulator's responsibilities and the facility's operational mission and should be based on trust and transparency. From a facility standpoint, the development and sustainment of this type of regulatory relationship is accomplished through open communication with regulators, making preparations for and accommodating inspections, providing access to records, assessments, test results, etc., and actively engaging with the regulator in a constructive manner. Based on my experience as both a regulator and a compliance manager, the alternative approach of being adversarial and evading active engagement with regulators does not lend itself to a successful and sustainable compliance strategy. Sustaining a positive and productive working relationship with a regulator is much easier if there is confidence that facility personnel arenot attempting to circumvent regulations, hide compliance issues or create intentional hindrances to regulatory oversight. Of course, a successful relationship depends on the regulators as well and the reciprocation of open communication, respect and transparency. A regulator or agency's scope of authority is bound by the laws and statutes that define an agency's regulatory role, and the expectation is that regulators/inspectors will conduct their activities within the scope of their authority. This is where it is important for facility compliance personnel to have an in-depth understanding of the regulations and the ability to utilize their regulatory expertise to work with the regulators on areas of disagreement regarding regulatory scope and/or interpretations. In a well-developed relationship, disagreements with an inspector or regulator can openly be discussed in a respectful manner and are most often successfully resolved in the facility's written responses to an inspection report, regulatory finding or notice of violation and not by engaging in a heated argument or becoming confrontational. Not all regulatory relationships are as easy and straightforward as conceptually presented here, but it is in the best interest of a facility owner/operator or compliance manager to develop an in-depth understanding of the environmental regulations that apply to their activities and develop a regulatory relationship strategy that fosters active engagement, open communication, transparency and mutual respect.

3.8   OTHER DIRECTIVES AND CORPORATE POLICY

Beyond the federal state and local regulations, there may be other directives or corporate policies that set forth requirements or otherwise influence a facility's environmental compliance programs and strategies. As an example, the NEPA regulations discussed earlier in the chapter require that federal departments and agencies develop procedures describing how the agency will implement the requirements established under NEPA for evaluating environmental impacts. These agency-level procedures define the set of requirements that apply to organizations within that specific agency. In other words, the procedural requirements for evaluating environmental impacts for projects proposed by the Department of Energy are different than those for the Bureau of Land Management. While both agencies are required to meet the requirements of NEPA, their implementation strategies and associated internal procedural requirements and directives are agency-specific. This can often be the case for federal facilities where federal agencies are commonly required to develop agency-specific procedures and directives that describe how organizations within the agency will meet the applicable regulatory requirements. These agency-specific procedures and directives often stipulate additional requirements such as establishing performance metrics, internal agency reporting, record-keeping, training and qualification.

In a similar fashion, corporations often utilize corporate procedures and/or policies as a means for implementing environmental regulatory and compliance programs within the corporation. Corporate directives are often used to implement programs within the organization for the purpose of minimizing hazardous chemical use, reducing waste, assessing compliance, tracking and correcting deficiencies and other activities focused on enhancing environmental performance, cost-effectiveness and the corporate reputation. Corporate policies and other directives are usually well communicated within the organization and do not pose the same challenges of determining applicability that can be associated with federal, state and local environmental regulations. However, requirements from corporate policies and/or agency directives are important elements requiring consideration and integration into a successful and sustainable compliance strategy.

3.9 ESTABLISHING AND DOCUMENTING APPLICABILITY

Establishing regulatory applicability can be accomplished through a variety of means. A viable approach involves several factors, including facility assessments to evaluate regulatory applicability and compliance, researching regulations, utilizing tools available from regulatory agency websites, actively engaging with regulatory agencies, baselining your activities with other facilities conducting similar activities and attending regulatory training classes and seminars. Many companies will use the services of an environmental consultant to conduct an assessment of their facility to assist in determining regulatory applicability. This approach is often beneficial and more cost-effective for companies that do not have in-house regulatory expertise. Companies that do have in-house environmental staff can conduct these assessments utilizing personnel with regulatory experience and subject matter knowledge. Regulatory applicability can and should be verified with regulatory agencies when it involves areas of the regulations that are interpretive and especially those that could place the company or facility in a situation leading to regulatory enforcement. This becomes an important aspect of the active engagement and developing regulatory relationships as discussed above.

It is important that personnel responsible for facility compliance fully understand the operations and activities occurring at the facility and which aspect of the operations are regulated and the reason they are regulated. Facilities that utilize an outside consultant to perform these assessments and make regulatory determinations should ensure that facility representatives participate in these assessments whenever possible. Facilities should require that the consultant provide a detailed description of their assessment findings to include the operations/activities that are regulated, which regulations apply, a discussion on why the regulations are applicable, the appropriate regulatory agency, any limitations of the assessment and any information relevant to implementing and maintaining compliance. This level of documentation is necessary to not only determine regulatory applicability and current compliance status but also in developing a strategy for sustaining environmental compliance in the future.

Facilities that have in-house environmental staff with regulatory expertise can not only utilize these staff members to assess operations and make regulatory determinations but they can also develop or participate in the development of regulatory applicability documentation. Regulatory applicability documentation should include the same information as identified above, including applicable regulations, specific operations/activities that are regulated, the reason specific operations are regulated and the regulatory agency point of contact. Again, documenting this level of detail will be beneficial in developing a compliance strategy as it will serve as a baseline document of applicable requirements that can be updated as operational activities change and will be instrumental in communicating and flowing down requirements within the organization.

Regulatory applicability determinations should be reassessed on a periodic basis to ensure they are still valid. There are numerous factors that could influence the frequency of these reassessments, including the complexity of operational activities, frequency of operational changes and new or revised regulatory requirements. Documentation of regulatory applicability should be managed and maintained as a company record. Because this documentation will serve as a focal aspect of a successful compliance program, it is important to ensure that the documentation is accessible, reproducible and that there is a mechanism for version control so that it is clear which version is current and when older versions were applicable. This does not necessarily have to be a complicated process and in most cases is as simple as identifying the documentation as a company record to be managed following established company records management procedures. Later chapters of this book will provide discussions relative to developing processes for using the information documented in regulatory applicability determinations to communicate and flow down requirements within the organization.

3.10 APPLIED LEARNING

Review the case study and respond to the proceeding questions.

3.10.1 CASE STUDY

Company ABC is responsible for a construction project that involves constructing a large warehouse building that includes installation of underground electrical utilities, a sanitary sewer system that connects to the local POTW and a large parking lot with storm drains. The building will require a special external metallic epoxy coating that results in a leftover residue that requires disposal as waste.

1. As the manager responsible for environmental compliance for this project, what environmental regulations need to be considered for applicability?
2. What resources might you use to assist in determining regulatory applicability?
3. Describe why you think specific regulations apply to this project.
4. For each regulation that is applicable, describe how you would determine the appropriate regulatory agency.

REFERENCES

ADEM. (2019). *NPDES permits.* 11 24. Accessed November 24, 2019. www.adem.state.al.us.

Broughton, Edward. (2005). The Bhopal disaster and its aftermath: A review. National Center for Biotechnology Information. June 7. Accessed December 12, 2019. www.ncbi.nlm.ni h.gov/pmc/articles/PMC114233/#!po=78.0702.

Browning, Randy. (2019). *Laqcey Act.* 11 14. Accessed November 15, 2019. www.fws.gov/int ernational/laws-treaties-agreements/us-conservation-laws/lacey act.

CEQ. (2019). *NEPA.GOV NAtional Environmental Policy Act.* December 13. Accessed December 13, 2019. www.ceq.doe.gov/laws-regulations/laws.

Copeland, Claudia. (2016). *Clean Water Act: A summary of the law.* Washington, DC: Congressional Research Service.

Ganzel, Bill. (2019). *Farming in the 1940s.* November 14. Accessed November 14, 2019. https ://livinghistoryfarm.org/farminginthe40s.

NDEP. (2019). *Stormwater discharge permits.* 11 24. Accessed November 24, 2019. www. ndep.nv.gov/water.

NOAA. (2019). *Critical habitat.* December 2. Accessed December 14, 2019. www.fisheries.n oaa.gov/national/endangered-species-conservation/critical-habitat.

Robinson, Jr., Eugene H. (2019). *Solid Waste Disposal Act (1965).* October 27. Accessed December 10, 2019. www.encyclopedia.com/histroy/encyclopedias-almanacs-transcri pts-and-maps-/solid-waste-disposal-act-1965.

U.S. EPA (2019). 40 CFR, part 122.2. *Code of Federal Regulations.* Washington, DC: Government Publishing Office, November 15.

U.S. Department of Commerce. (2019). *U.S. and world population clock.* 11 15. Accessed November 15, 2019. www.census.gov.

U.S. EPA. (2004). *Understanding the Safe Drinking Water Act.* U.S. EPA. June. Accessed December 13, 2019. www.epa.gov/sites/production/files/2015-04/documents/ epa816fo4030.

U.S. Department of Commerce. (2010). *A facility owner/operator's guide to oil pollution pre-vention.* U.S. EPA. June. Accessed December 1, 2019. www.epa.gov/sites/production/f iles/documents/spccbluebroch.

U.S. Department of Commerce. (2016). *1990 Clean Air Act amendment summary: Title V.* October 19. Accessed December 7, 2019. www.epa.gov/clean-air-act-overview/1990-clean-air-act-summary-title-v.

U.S. Department of Commerce. (2016). *Criteria air pollutants.* December 20. Accessed December 5, 2019. www.epa.gov/criteria-air-pollutants.

U.S. Department of Commerce. (2017). *Evolution of the Clean Air Act.* 1 3. Accessed December 4, 2019. www.epa.gov/clean-air-act-overview/evolution-clean-air-act.

U.S. Department of Commerce. (2017). *Learn about underground storage tanks (USTs).* April 11. Accessed December 11, 2019. www.epa.gov/ust/learn-about-underground-stora ge-tanks-usts.

U.S. Department of Commerce. (2017). *Mobile Sources Compliance Monitoring Program.* February 8. Accessed December 6, 2019. www.epa.gov/compliance/mobile sources-compliance-monitoring-program.

U.S. Department of Commerce. (2017). *National Environmental Policy Act review process.* January 24. Accessed December 13, 2019. www.epa.gov/nepa/national-environmental-policy-act-review-process.

U.S. Department of Commerce. (2017). *Reducing emissions of hazardous air pollutants.* February 9. Accessed December 5, 2019. www.epa.gov/haps/reducing-air-emissions-hazardous-air-pollutants.

U.S. Department of Commerce. (2017). *Safe Drinking Water Act.* January 12. Accessed December 13, 2019. www.epa.gov/sdwa.

U.S. Department of Commerce. (2018). *About the Endangered Species Protection Program.* May 29. Accessed December 14, 2019. https://www.epa.gov/endangered-species/about-endangered-species-protection-program.

U.S. Department of Commerce. (2018). *Facility response plan applicability.* 4 4. Accessed December 3, 2019. www.epa.gov/oil-spills-prevention-and-preparedness-regulations/fa cility-response-plan-frp.

U.S. Department of Commerce. (2018). *Key elements to include in a facility response plan.* 4 4. Accessed December 3, 2019. www.epa.gov/oil-spills-prevention-and preparedness-re gulations/key-elements-include-facility-response-plan-frp.

U.S. Department of Commerce. (2018). *Phase-out of ozone-depleting substances.* November 28. Accessed December 8, 2019. www.epa.gov/ods-phaseout.

U.S. Department of Commerce. (2018). *Superfund: CERCLA overview.* June 4. Accessed December 16, 2019. www.epa.gov/superfund/superfund-cercla-overview.

U.S. Department of Commerce. (2019). 40 CFR 122.23. *Code of Federal Regulations.* Washington, DC: Government Publishing Office, November 25.

U.S. Department of Commerce. (2019). 40 CFR Part 264 subpart B. *Code of Federal Regulations.* Washington, DC: Government Publishing Office, November 21.

U.S. Department of Commerce. (2019). 40 CFR part 110. *Code of Federal Regulations.* Washington, DC: US Government Publishing Office, November 26.

U.S. Department of Commerce. (2019). 40 CFR part 112.6. *Code of Federal Regulations.* Washington, DC: US Government Publishing Office, November 26.

U.S. Department of Commerce. (2019). *Background about compensatory mitigation require-ments under CWA Section 404.* 4 16. Accessed December 3, 2019. www.epa.gov/cwa -404/background-about-comensatory-mitigation-requirements-under-cwa-section-404.

U.S. Department of Commerce. (2019). *Clean Air Act title IV - noise pollution.* August 5. Accessed December 6, 2019. www.epa.gov/clean-air-act/clean-air-act-title -iv)_noise-pollution.

U.S. Department of Commerce. (2019). *Federal Insecticide, Fungicide, and Rodenticide Act (FIFRA).* June 12. Accessed December 14, 2019. www.epa.gov/enforcement/federal-ins ecticide-fungicide-and-rodenticide-act-fifra-and-federal-facilities.

U.S. Department of Commerce. (2019). *Final 2015 MSGP documents.* April 12. Accessed November 25, 2019. www.epa.gov/npdes/final-2015-msgp-documents.

U.S. Department of Commerce. (2019). *Lead.* November 14. Accessed December 12, 2019. www.epa.gov/lead.

U.S. Department of Commerce. (2019). *Managing used oil: answers to frequent questions for business.* April 29. Accessed December 11, 2019. www.epa.gov/hw/managing-used-o il-answers-frequent-questions-business.

U.S. Department of Commerce. (2019). *Mechanisms for providing compensatory mitigation under CWA Section 404*. 4 16. Accessed December 3, 2019. www.epa.gov/cwa-404/mechanisms-providing-compensatory-mitigation-under-cwa-section-404.

U.S. Department of Commerce. (2019). *National Pretreatment Program*. November 5. Accessed November 20, 2019. www.epa.gov/npdes/national-treatment program.

U.S. Department of Commerce. (2019). *NPDES Stormwater Program*. August 12. Accessed November 24, 2019. www.epa.gov/npdes/npdes-stormwater-program.

U.S. Department of Commerce. (2019). *Permit program under CWA section 404*. 5 14. Accessed December 3, 2019. www.epa.gov/cwa-404/permit-program-under-cwa-section-404.

U.S. Department of Commerce. (2019). *Summary of the Clean Water Act*. 11 18. Accessed November 15, 2019. www.epa.gov/laws-regulations/summary-clean-water-act.

U.S. Department of Commerce. (2019). *Stormwater discharges from municipal sources*. September 11. Accessed November 22, 2019. www.epa.gov/npdes/stromwater/municipal-sources.

U.S. Department of Commerce. (2019). *Universal waste*. October 21. Accessed December 11, 2019. www.epa.gov/hw/universal-waste#state.

U.S. Department of Commerce. (2019). *What is toxics release inventory*. September 27. Accessed December 12, 2019. www.epa.gov/toxics-release-inventory-tri-program.

U.S. Department of Commerce. (2019). *What treatment standards apply to my waste*. November 2019. Accessed December 11, 2019. www.epa.gov/hw/treatment-standrads-waste-subject-land-disposal-restrictions#apply.

United States Congress. (1977). *42 U.S.C. Title 42*. Washington, DC, August 7.

United States Congress. (1996). *H.R.2036 - Land disposal Program Flexibility Act of 1996*. March 26. Accessed December 10, 2019. www.congress.gov/bill/104th-congress/house-bill/2036.

United States FWS. (2013). *ESA basics*. fws.gov. January. Accessed December 14, 2019. www.fws.gov/endangered/esa-library/pdf/ESA_basics.

United States FWS. (2018). *Endangered species*. December 11. Accessed December 14, 2019. www.fws.gov/endangered/laws-policies/index.html.

4 Implementation Roles, Responsibilities and Accountability

4.1 LEADERSHIP CHARTING THE WAY

A successful environmental compliance program in the corporate world today requires a level of commitment by management that promotes and supports the integration of compliance into the corporate culture. The attitude taken by corporate managers and leadership regarding environmental compliance will set the tone for the organization as a whole and have a direct bearing on the level of importance employees place on compliance in the performance of their job duties. Corporations that integrate environmental compliance into their strategic planning and daily operations must also clearly communicate their vision and expectations throughout the organizational structure. An environmental compliance policy that is openly promoted by management and consistently emphasized is much more likely to be embraced by employees as an important aspect of their job responsibilities rather than just a necessary inconvenience.

Empowering employees to take ownership of this type of corporate priority is more readily accommodated by a corporate structure which includes a process for governing the flow down and implementation of environmental regulatory requirements. As depicted in Figure 4.1, a business model that provides for a proceduralized process to identify and flow down regulatory requirements creates opportunities for employees to participate and can also serve as an invitation for them to take on responsibilities within the process. An established process for requirements flow down will not only provide a means for empowering employees, it will also ensure requirements are consistently flowed down throughout the organization. Corporate environmental policies that are clearly communicated and consistently supported by all levels of management not only help to emphasize the importance of the policies, but can also serve as a strong catalyst in creating a positive and impassioned culture throughout the organization.

4.1.1 LEADERSHIP STYLE AND IMPACT

There are many different leadership styles applied in business and industry, and each has been labeled based on the general characteristics of the associated style of management (Figure 4.2). Some of the more common leadership styles often found in academic and instructional literature include:

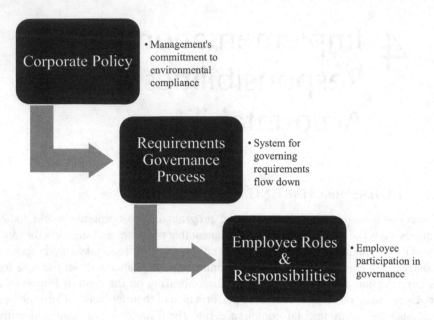

FIGURE 4.1   Governing requirements flow down.

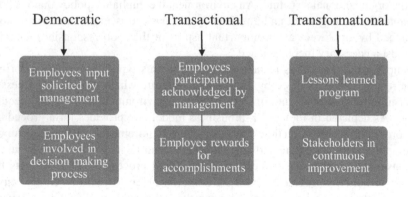

FIGURE 4.2   Leadership styles and employee participation.

- Transformational – always transforming, emphasizes continuous improvement
- Transactional – focused on achieving or exceeding established goals
- Democratic – employee input solicited in decision-making
- Autocratic – does not solicit employee input in decision-making
- *Laissez-faire* – hands-off approach, employees are entrusted to get their work done with limited management oversight
- Bureaucratic – focused on following established rules, regulations and policies
- Situational – utilizes a mix of leadership styles, depending on the situation

Each style of leadership has a characteristic that defines the style, and each most likely has specific environments where their focus may be more conducive to a desired result. Two questions immediately come to mind when considering leadership styles and environmental compliance: Which leadership style is best suited for organizations responsible for environmental compliance? Is a single leadership style adequate or does utilizing various combinations of leadership styles produce better results?

A study on leadership styles in the North American environmental sector found that leaders in the environmental sector often utilize a mix of leadership styles and due to the dynamic and complex nature of the environmental field, environmental managers need a diverse set of leadership skills and the ability to rapidly change between styles. The same study also proposes a model where leadership skills are associated with an ecocentric set of values and a focus on creating and maintaining sustainable relationships with employees and other stakeholders (Egri and Herman 2000). The findings of this study clearly propose that utilizing a single leadership style would not be optimum for environmental leaders. Considering the complexity of environmental compliance, it is reasonable to conclude that environmental compliance managers would benefit from possessing a well-rounded set of leadership skills that encompass aspects from numerous leadership styles, and this is best expressed as a situational leadership style.

The size of a company will obviously impact the organizational structure and associated management structure. Larger companies with more complex organizational structures inherently create a need for a greater number of managers as compared to smaller companies operating with fewer employees. The organizational mission and goals of a company will undoubtedly also exhibit influence on managers and their approaches for leadership. However, one can argue that any company regardless of size and mission can benefit from managers who possess situational leadership skills and are well versed in utilizing the collective skill set. Each of the leadership styles described above is typified by a characteristic that can be beneficial to an organization. The importance of continuous improvement, employee recognition, democratic decision-making, a delegatory management style and adherence to the rules all have their place in business and industry. Companies that operate in highly regulated environments can benefit from a positive and proactive compliance culture with a foundation built on empowering employees through the utilization of a situational leadership style that emphasizes employee involvement, recognition, process ownership and responsibility.

4.1.2 COMMUNICATING THE IMPORTANCE OF ENVIRONMENTAL COMPLIANCE

One of the most important aspects of a successful compliance program is a commitment by management to support the integration of compliance into business activities and operations. As discussed in the above section on leadership styles, creating a proactive compliance culture must begin at the top levels of management with a sincere and committed objective of enabling the desired culture to flourish. It is important for employees to fully understand and appreciate managements' attitude toward environmental compliance. A one-time declaration by management on the importance of environmental compliance is far less effective than a documented policy or mission statement that is visible, accessible and its importance continually reiterated by management. In regard to organizational communication, top-down communication is an important aspect that when used effectively can result in company policies having more of a positive influence on employee attitudes and conduct (Guest and Conway 2006).

However, communicating the importance of environmental compliance goes beyond documenting and reiterating a company's environmental policy. The management commitment to environmental compliance also has to be integrated into and emphasized on other ongoing organization elements such as training, requirements governance, work procedures, process improvement, stakeholder relationships, public affairs messaging and commitment of resources. The employees' perception of a company's level of commitment to environmental compliance is based on the degree that management leads and actively supports the integration of compliance into relevant aspects of the organization. A company that makes the statement that environmental compliance is important but fails to integrate it into business practices, processes and training is in reality sending a diluted message to employees that will not encourage a proactive culture and most likely will result in a more indolent attitude toward compliance.

4.2   EMPLOYEE PARTICIPATION

Arguably the most important element in building a sustainable compliance culture is active employee participation. As discussed in the previous section, employees tend to place a level of enthusiasm on aspects of work that are consistent with their perception of how important that same aspect is to management. In other words, employees will follow management's lead, and where corporate policy, management systems and leadership styles promote and enable participation, employees are more inclined to take the initiative to actively get involved. When this type of proactive approach is presented, it creates an environment that can motivate employees to readily embrace management's position and take ownership of their compliance responsibilities. A requisite component of a company's strategy for a sustainable environmental compliance program is one that emphasizes employee participation. (Gollan and Xu 2015) propose the concept that employee participation very well could become a necessity for companies striving to create a sustainable organization.

The importance of a situational leadership style that utilizes a range of styles can be highlighted here to emphasize how the various styles promote employee's participation.

Employee participation is critical to growing and sustaining a proactive compliance culture. An organization that can successfully establish active employee participation places the organization in the position where a proactive compliance culture can become a grassroots effort and evolve into the recognized corporate mindset.

4.2.1 WHERE IMPLEMENTATION REALLY TAKES PLACE

Businesses that operate in highly regulated environments are inherently faced with the decision on how to best align business needs with regulatory compliance. Business approaches can range from one that operates with a 'don't get caught' mentality and the assumption of avoidant or confrontational relationships with regulators to one that operates compliantly, has developed a constructive relationship with regulators and successfully meet business objectives. Companies that have established sustainable environmental compliance programs have done so by successfully integrating regulatory compliance into their business practices and operational processes throughout all levels of the organization. Operating successfully in highly regulated environments requires the identification of applicable regulatory requirements, flowing those requirements down to the relevant parts of the organization, integrating applicable requirements into work processes and finally implementing the requirements during the performance of work activities. Consistent implementation of regulatory requirements during the performance of work is fundamental to the success of the overall compliance program. A company can have established processes for identifying and flowing down regulatory requirements, but if these regulations are not adequately implemented, the company is placed in noncompliant situation and vulnerable to regulatory violations and penalties.

Here again is an example of just how crucial of a role employee participation plays in a compliance program. Employees tasked with actually implementing regulatory requirements as part of their work assignments can often carry a significant responsibility for ensuring compliance with the requirements is achieved. The employee's ability to meet these compliance responsibilities are enhanced within a proactive compliance culture where management solicits employee input regarding their work processes and takes deliberate actions to improve them as needed. Work processes that are performed using clear and unambiguous instructions with well-defined roles and responsibilities tend to consistently produce the intended result or product. Organizational leaders and managers help improve their employees' ability to meet their responsibilities for regulatory compliance when these types of work controls are made available and supported. These types of resources not only make the employee's job easier, but also provide for consistent results, which in this case is critical since regulatory compliance can be a significant impact on the company's bottom line.

Companies that proactively integrate environmental compliance into their organization upfront better position themselves to realize a lower long-term cost of compliance as compared to companies that take more of a reactive approach. The cost to implement a regulatory compliance program is obviously dependent on the size of the company and the complexity of the applicable regulations. However, the cost of noncompliance often involves more than simply the dollar amount of an assessed penalty or fine. Impacts of noncompliance that are often overlooked include:

- Corporate liability
- Personal liability
- Damaged corporate reputation
- Legal fees

- Reduced business opportunities
- Costs for cleaning up releases
- Loss of employees

These often unanticipated impacts can equate to significant costs and financial burdens for companies. The potential for a company to be impacted by these costs of noncompliance is greatly reduced with an integrated environmental compliance system that recognizes the importance of implementing and meeting regulatory requirements at the work activity level.

4.2.2 Motivated versus Reluctant Implementation

Another significant factor that influences how well employees perform their jobs is their level of motivation. Employers obviously desire employees who are motivated to perform their duties in a manner that meets or even exceeds the expectations for the job. Compliance managers can often face challenges with employee motivation because it is frequently the case that responsibilities for meeting environmental regulatory requirements are added responsibilities to an employee's core set of job duties. An employee with regulatory compliance responsibilities who is motivated to perform at a high level and strives to ensure regulatory requirements are adequately addressed is an obvious preference over an employee who is less enthused and reluctantly performs these types of additional job tasks.

Thus far this chapter has included somewhat of a top-down discussion regarding conceptual elements of a strategic compliance strategy that can promote and encourage active employee participation (Figure 4.3). The argument can be made

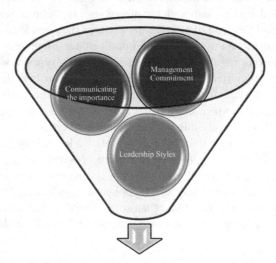

Employee Participation

FIGURE 4.3 Elements that promote employee participation.

that there is a strong correlation between employee participation and employee motivation in the workplace. Workplace environments which provide the elements supportive of active employee participation also experience positive levels of employee job satisfaction. Companies where employee participation in work decisions is associated with a significant amount of influence on the work itself experienced relatively consistent and positive outcomes relative to productivity and employee satisfaction (Cotton et al. 1988).

The challenge of motivating employees to consistently perform at high levels can be diminished when employees experience higher levels of job satisfaction. Obviously, rewarding performance also adds to employee satisfaction, but there is a contrary side that must also be considered by managers when considering employee motivation. If the employees who perform their compliance duties reluctantly and at substandard levels are rewarded similarly to the motivated high-performing employees, continuance of the reluctant employees' substandard performance is enabled. The effects of this type of acknowledgment and reward can foster an environment that leads to decreased employee motivation which when left unabated can also contribute to a decreased desire by employees to actively participate in work processes.

There are various other actions that can influence employee motivation and possibly even enable reluctant behavior toward implementation of regulatory requirements. Oftentimes the actions or attitudes being propagated within the organization are unintentional, but can still result in unfavorable results in regard to promoting motivated implementation. Table 4.1 presents a list of management actions and the potential impacts on employee motivation.

TABLE 4.1
Management Actions and Impacts on Employee Motivation

Management Action	Yes	No
Commitment to compliance	Employees feel supported by management	Perceived lack of support by management
Communicates importance of compliance	Employees feel valued in their role	Employees contribution to mission perceived as less valued
Positive perception of compliance relative to corporate mission	Employees feel they are contributing to success of company	Employees compliance role perceived as barrier to corporate mission
Supports integration of compliance into organizational elements	Employees feel supported by management	Employees feel organizationally isolated
Adequate resources prioritized	Increased employee morale, higher employee retention	High workloads and employee burn-out
Employee participation promoted	Increased job satisfaction, employee ownership in work processes	Autocratic environment. No sense of ownership and lower job satisfaction
Accomplishments acknowledged	Employees feel appreciated and valued	Employees feel unappreciated and that their efforts are unrecognized

The discussion in this section is intended to emphasize the connection between active employee participation and the level of employee motivation when implementing regulatory requirements. Both are essential components in creating a sustainable compliance organization, and compliance managers should exercise vigilance in their methods for fostering participation and motivating employees.

4.3 ENVIRONMENTAL SUBJECT MATTER EXPERTS AND PROFESSIONALS

Successful strategies for environmental compliance are often built around an organization designed to identify regulatory requirements, determine applicability and facilitate implementation. Companies that incorporate regulatory subject matter experts (SMEs) roles into their compliance organization create an environment that promotes integration of regulatory compliance into the relevant business units as required. The role of environmental professionals and SMEs can provide for consistency in regulatory evaluations, applicability determinations and valuable regulatory expertise to organization elements and business units. Incorporating the SME roles into environmental compliance strategies allows for other organizational elements within the company to maintain a focus on their primary mission while relying on the environmental professionals and SMEs to provide regulatory expertise and effective implementation strategies. The challenge often presented by regulatory compliance is finding a synergetic balance between the corporate or organizational mission and implementing regulatory requirements. This is where the concept of integrating environmental SMEs into the organization can significantly mitigate many of the difficulties in achieving the desired balance.

Of course, the number of SMEs and environmental professionals employed by a company will be dictated by the size of the company, regulatory complexities and available resources. However, regardless of the size of the company, the concept of the SME fulfilling the role of regulatory expert and the established point-of-contact for environmental compliance remains the same. Figures 4.4 and 4.5 provide two conceptual examples of the organizational integration of SMEs into various sized organizations.

Figure 4.4 shows a simplified example of how a larger sized company with a more diverse organization might integrate environmental professionals and SMEs

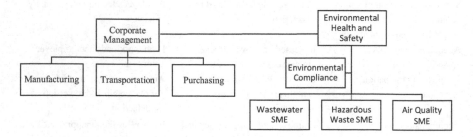

FIGURE 4.4 Integration of environmental SMEs into large organization.

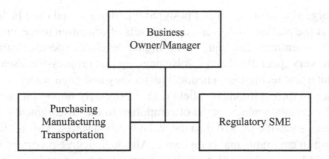

FIGURE 4.5   Integration of environmental SME into small company.

into the organization. Large companies typically have the resources to employ larger numbers of employees and often have the ability to create an environmental compliance group or division staffed by multiple environmental professionals. Highly regulated operations can benefit from having multiple environmental professionals with subject matter expertise on regulations that specifically apply to the company. In this example, each SME is responsible for a specific regulatory area such as wastewater, air quality or hazardous waste. Figure 4.5 depicts a smaller company with fewer employees that may not have the resources to employ multiple environmental professionals to support regulatory compliance. In this case, environmental compliance may be the responsibility of a single individual who is responsible for all areas of environmental regulations. Again, the number of environmental professionals employed by a company will be dependent on various factors, and there are certainly other organizational approaches that can be considered for environmental compliance. The point to be made here is that incorporating environmental SMEs into the organizational structure is an imperative element for any strategy aimed at obtaining a sustainable compliance program.

4.3.1   What Defines an Environmental Subject Matter Expert?

Webster defines the term 'expert' as one having, involving or displaying special skill or knowledge derived from training or experience (Merriam-Webster n.d.). In the environmental compliance realm, a subject matter expert is a person that has genuine expert knowledge regarding what it takes to successfully identify, understand and implement environmental regulations. Environmental SMEs exhibit high levels of expertise related to a specific subject area such as hazardous waste or to a collection of areas within the environmental field. The three primary components that contribute to an SME's expertise are:

*   Education
*   Experience
*   Training

An SME's expertise may be attributable to any combination of these three elements as each of them has a definite importance in establishing expertise. Education can

include undergraduate, graduate and postgraduate degrees and even include technical training at the trade school level. Higher levels of education in the environmental
field of study oftentimes can equate to a higher level of expertise, particularly in
areas that are very specialized and involve ongoing and progressive research such as
pollution abatement technology, chemical toxicology and environmental case law.

Experience in the environmental field is also extremely important in establishing
expertise. The common adage that is often applied to the environmental field is that
'your degree gets you the job and then the real education starts.' This is often the case
when working in environmental compliance. Although degree programs in the environmental field have evolved substantially over the last two decades, it is impossible
to provide a curriculum that can adequately address the infinite number of compliance situations and regulatory applications that are presented in business and industry
today. As with education, a person's experience can contribute greatly to their level
of expertise. As an example, there are a lot of regulatory SMEs currently working
in environmental compliance for private companies who got their start working for
regulatory agencies. SMEs with this type of experience in their background can be
invaluable to compliance organizations because they not only possess an in-depth
knowledge of the regulations, but they also have a genuine understanding of how regulatory agencies assess regulatory compliance. Likewise, a person who has worked
in compliance in private industry for many years has most likely been involved in
numerous compliance scenarios and acquired a great deal of regulatory experience.

Training plays two important roles in regard to environmental SMEs. First,
there is a need for training to cover a level of specificity that cannot reasonably be
addressed by a college degree curriculum. Examples include air permitting for specific applications, closing underground storage tanks, storm water management and
environmental compliance auditing, to name but a few. Second, there is a need for
SMEs to maintain their level of expertise and competence in a field that is typified by
constantly changing regulations, new methods for achieving compliance, modified
legal interpretations and new technologies.

Again, there is no single combination of education, experience and training that is
the standard for defining what constitutes being recognized as an SME. The necessary
levels of education, experience and training are really circumstantially dependent, and
companies must consider the level of knowledge necessary to accomplish environmental compliance for their specific situation. Companies typically approach this by
stipulating the required levels of education, experience and training for specific job
positions and are commonly emphasized in positions that include an SME role.

4.3.2　Training, Experience and Keeping Skills Relevant

Training, continuing education and ongoing relevant experience are all instrumental
in maintaining a relevant skill set for SMEs. The environmental compliance field
involves a multitiered regulatory model that includes regulatory requirements at the
federal, state and local levels. Part of being an environmental SME is staying current and keeping up with regulatory changes. While there is an overall parallelism
in regulations as they flow down from the federal level, there are commonly nuances

and variations to the regulations at the state and local levels. Environmental regulations are constantly changing at all levels, and keeping up with new regulations and regulatory changes is paramount to the role of an environmental SME. This can be accomplished through training, continuing education courses, attending regulatory workshops, subscription to regulatory publications, tracking environmental legislation and participating in industry-sponsored working groups.

Continuing relevant experience is important for SME skill sets, and they can benefit from expanding their areas of experience to include other areas of the environmental field whenever possible. There is a substantial amount of connectivity between the various aspects of the environmental field, and this is particularly true in the case of environmental regulations and compliance. The broader the SMEs' understanding of the environmental regulations are, the better equipped they are to develop complete and thorough compliance implementation strategies. However, due to the vast number of environmental regulations in place today and the associated complexities, it is unrealistic to expect that any one person could reasonably be an SME on every environmental law and the entire collection of regulations. A company whose operations warrant compliance to one or two regulations may require only one SME for successful compliance. However, companies that are required to comply with a large number of environmental regulations will benefit from having multiple SMEs. In this case, it is important to balance available resources and regulatory-related duties in order to promote the ability of personnel to maintain subject matter expertise and not be forced into operating as more of a generalist due to unreasonable workloads. Again, this will be driven by the size of the company, the number and complexity of regulatory obligations the company is responsible for and available resources. In all cases, SMEs should continue to be involved in work activities that provide relevant work experience in order to maintain their skill set.

Depending on the type of industry, it may also be important for an SME to stay current with innovations and changes in technologies. Technology is increasingly relied upon in today's corporate environment, and its prominence in industry and business requires a focused attention to keep up with the rapid pace of change and innovation. Technology applications in the environmental field can include pollution abatement, contaminant monitoring, release detection, exposure monitoring as well as a plethora of environmental compliance computer applications and software. Keeping abreast of the technologies that can have a beneficial application to the company is critical for the SME and appreciable effort should be applied in this regard. Methods that can be employed in this regard include vendor-sponsored training, regulatory workshops and networking with peers from similar industries.

4.3.3 COMPLIANCE DECISION-MAKING

Identifying environmental regulations and understanding how they apply to specific activities or operations are the first important steps toward successful regulatory compliance. Companies can benefit from instituting processes that help ensure correct identification of regulations and a complete understanding of how the regulatory requirements apply to their specific operations. It is important that the evaluation of regulatory applicability be conducted consistently and thoroughly in order to achieve effective and compliant implementation. This reiterates the importance of the SME role and the establishment of a documented process for identifying and governing the flow down of applicable regulatory requirements to the appropriate organizational elements. Figure 4.6 provides a flow diagram that depicts the rudimentary elements and participants of a conceptual requirements identification and governance process that can benefit compliance decision-making.

While this simplified flowchart depicts a conceptual approach for internal compliance decision-making, there are other steps that should be considered for integration into the decision-making process. Networking with colleagues and peers working within similar industries can provide valuable insight into compliance decision-making. Chances are that other companies are faced with the same or similar compliance decisions and may be able to share some suggested strategies and useful lessons learned. This is also where having a positive working relationship with regulators can pay dividends. The ability to solicit input from the regulators and openly discuss implementation strategies can be invaluable. Regulators have the benefit of being able to observe many different approaches taken by companies regarding implementation of specific regulatory requirements. As such, they can provide insight as to which approaches are acceptable from the regulators' standpoint and which ones should be avoided. Information gained from networking with industry peers and regulatory interactions can greatly contribute to internal compliance decision-making and, as such, should be integral elements of a company's decision-making process.

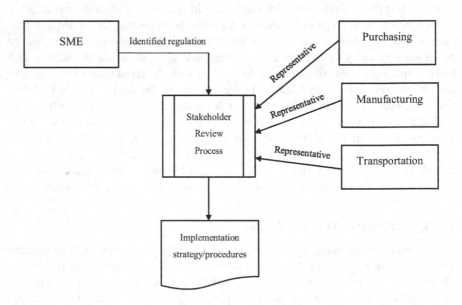

FIGURE 4.6   Compliance decision flowchart.

4.4   REMOVING AMBIGUITY FROM COMPLIANCE DECISIONS

There can be an enormous amount of ambiguity in the applicability of environmental regulations, and environmental professionals are often faced with the challenge of deciphering how regulatory requirements apply to their specific operations. Ambiguity may be a result of the way the regulation is written or of a lack of understanding regarding the specifics of operations, or a combination of the two. The point here is that while some regulations are clear and concise, there are plenty of gray areas within the regulations that can make compliance decisions challenging. The better a company is at removing ambiguity from their compliance decisions, the more they can control the risk of noncompliance and the avoidance of regulatory violations, penalties and other costs associated with noncompliance. In my experience, the most valuable method for removing the ambiguity from compliance decisions has been the utilization of the integrated team approach. Integrated teams can be applied at the project level, division level, corporate level, institutional level, regional level or at any level where there are stakeholders or knowledgeable people who can contribute to the decision. Integrated teams ensure that compliance decisions are not being made in a vacuum without consideration being given to everyone that may be impacted by the decision or that may have information that is pertinent to the decision. Not surprisingly, the concept of using integrated teams is not unlike the stakeholder review presented in Section 4.3, with the only difference being the latter is presented as a corporate or an institutional process for governing requirements set at a specific organization level.

The integrated team approach can be used for one-time projects or ongoing operations which involve regulatory compliance decisions. Integrated teams help to ensure that all aspects of the project, operations, regulations and other relevant information are considered in the decision. The value of an integrated team is not limited to situations where the regulatory requirements may be vague. They also are instrumental in ensuring that the operational aspects of the project or activity are understood, including project scope, schedule, timelines, deliverables, associated equipment, materials, air emissions, anticipated waste products, etc., etc., etc. Integrated planning teams are commonly used at the project level to ensure that project planning adequately considers input from all stakeholders. However, integrated teams are also beneficial in decision-making for ongoing programs, especially those that are highly regulated. As an example, having an integrated program team that meets on an ongoing and regular basis to discuss potential regulatory or operational issues can also be extremely effective in removing ambiguity from compliance decisions and fostering continuity within ongoing programs or operations.

Integrated team membership does not necessarily have to be limited to company personnel and can frequently benefit from the inclusion of members from outside the organization such as personnel from similar companies, regulators, industry SMEs, subcontractors, vendors and program sponsors. Other outside information that may be useful to an integrated team's decision-making and shouldn't be ignored are sources such as documented lessons learned, regulatory workshops, conferences and workshops.

4.5   APPLIED LEARNING

Review the case study and respond to the questions that follow.

4.5.1   CASE STUDY

You are a newly hired environmental compliance manager for Company B, a chemical research and development company that also manufactures and distributes numerous chemical products. Company B has a good overall compliance record in all areas except the chemical research department. The research department staff don't seem to take environmental compliance very seriously and think all of these environmental requirements provide no added value to the company's mission.

1. In your role as compliance manager, what steps would you take to cultivate a more positive attitude toward compliance among the research department staff?
2. What do you think may be some of the leading causes that allowed this type of anti-compliance attitude to take hold in the research department?
3. In your opinion, what role should Company B management play in helping to transform the way in which the research department staff prioritize compliance?
4. How would you utilize environmental compliance department staff in addressing this situation?

REFERENCES

Cotton, John L., Vollrath, David A., Froggatt, Kirk L., Lengnick-Hall, Mark L., & Jennings, Kenneth R. (1988). Employee participation: Diverse forms and different outcomes. *The Academy of Management Review*, Vol. 13, No 1, pp. 8–22.

Egri, Carolyn P., & Herman, Susan. (2000). Leadership in the North American environmental sector: Values, leadership styles, and contexts of environmental leaders and their organizations. *The Academy of Management Journal*, Vol. 43, No. 4, pp. 571–604.

Gollan, Paul J., & Xu, Ying. (2015). Re-engagement with the employee participation debate: Beyond the case of contested and captured terrain. *Work, Employment & Society*, Vol. 29, No. 2, pp. NP1–NP13.

Guest, David E., & Conway, Neil. (2006). Communicating the psychological contract: An employer perspective. *Human Resource Management Journal*, Vol. 12, No.2, pp. 22–38.

Merriam-Webster. (n.d.). *The Merriam-Webster.com dictionary*. Accessed January 12, 2020. https://www.merriam-webster.com/dictionary/expert.

5 Flow Down of Requirements

5.1 COMMUNICATING REQUIREMENTS WITHIN THE ORGANIZATION

Chapter 3 included discussion on the major environmental laws and regulations and the importance of identifying the regulatory requirements that are applicable to activities and processes within a corporation or business. The importance of correctly identifying applicable regulatory requirements cannot be overstated. It is one of the first critical steps in a successful compliance strategy and, as discussed earlier, it not only requires a focused upfront effort, but also diligence in keeping abreast of new and changing regulations. Once applicable regulatory requirements are identified, they must be communicated to the relevant parts of the organization in order to allow for implementation. Successful compliance programs include defined processes for communicating regulatory requirements within the organization. Communication strategies developed for information sharing should include an effective flow of concise information that is thorough and accessible (Carlile 2011). It is essential that the flow down of requirements within the corporate structure is accomplished using a defined process that ensures an accurate and consistent approach in communicating regulatory requirements and flowing them down to the appropriate organizational elements where they can be implemented. The concept of requirements flow down is that the information becomes more detailed and specific as it flows down from requirement documents to the corporate level, organizational level and finally to the work unit level. At the corporate level, the information is commonly documented at a level that identifies an environmental law or a set of regulations that apply to the company. As the information flows down to the organizational elements within the corporation, the information typically becomes more specific to include the affected activities or processes as well as the specific portions of the regulations that apply. When requirement flow down reaches the work unit level, it is most advantageous if the level of information includes details on the specific requirements as they apply to the work activity or process and the steps necessary to comply with the regulatory requirement (see Figure 5.1).

5.2 DEVELOPING PROCESSES FOR REQUIREMENTS FLOW DOWN

When developing a process for flowing down regulatory requirements, it is important to create a process that is foundational within the corporation or company. To be successful, the process must involve participation from all relevant aspects of

FIGURE 5.1   Requirement flow down information levels.

the organization that are affected by environmental regulations. Organizational elements can be affected by environmental regulations in many different ways:

* Management responsibility
* Implementation responsibility
* Oversight responsibility
* Environmental reporting responsibility
* Administrative (e.g., procedures, records, document control)
* Environmental training and qualifications

Each of the examples depicts areas of responsibilities that can play an important role in communicating environmental regulatory requirements and as such should be included as stakeholders in the development and function of a requirements flow down process.

The organizational elements that have aspects of environmental compliance responsibilities within a company or corporate structure can be utilized specifically in the process of governing and communicating requirements internally within the organization. Establishing specific roles and responsibilities for organizational elements with regard to environment-related requirements helps establish an organized framework for a requirements flow down process. Figure 5.2 shows an example of how organizational elements can be utilized for managing requirements identification and flow down.

In this example, each organizational element has assigned areas of regulatory responsibility that functionally fit within their organizational duties. The regulatory responsibilities would include identifying applicable requirements, keeping current on new regulations and regulatory updates and providing this information to the requirements flow down process. In many cases, the areas of regulatory responsibility can be established to mirror the organization of the company, but there is no hard and fast rule that applies here. Organizations should strive to establish regulatory areas of responsibility that best align with their organization structure and promote clear roles and responsibilities coupled with accurate and timely requirements identification and flow down. To ensure continuity within and among each area of

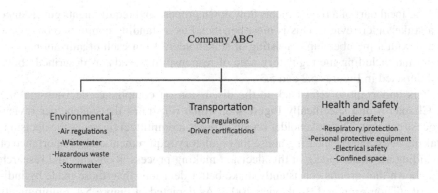

FIGURE 5.2 Areas of regulatory responsibility.

regulatory responsibility, it is beneficial to designate a manager for each regulatory responsible area. This position will serve as the lead for managing requirements identification and flow down for a specific set or sets of regulations. The importance of this role is twofold in that each area of regulatory responsibility has a designated point of contact that serves as a resource for regulatory information and also represents the organizational element as a participant in the requirements governance and flow down process.

Depending on the size and complexity of the environmental regulatory environment within the company, regulatory SMEs can be assigned to specific sets of regulations and support the management of regulatory requirements within the organizational area of responsibility. Figure 5.3 shows responsible area leaders and SMEs added to the areas of regulatory responsibility. The number of SMEs and associated assignments can be highly variable based on the size of the company and the availability of personnel and resources. The configuration depicted in the figure is an example to illustrate the concept only.

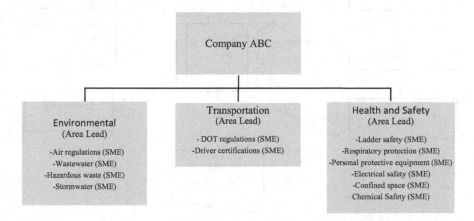

FIGURE 5.3 Areas of regulatory responsibility with leads and SMEs.

The focal part of a requirements flow down process and requirements governance is a stakeholder review. This is most beneficial as a standing committee or review board with a membership consisting of stakeholders from each organizational element and including the regulatory area of responsibility leads, as described above and depicted in Figures 5.2 and 5.3.

It is important to refer back to the discussion on compliance decision-making in Chapter 4 and specifically Figure 4.6, which illustrates the stakeholder review process concept. The stakeholder review process (committee) serves as the decision-making body and the clearing house for regulatory requirements. The importance of including all stakeholders in this decision-making process is obvious and research has shown that groups consistently make better decisions than those made by individuals (Thompson and Brajkovich 2003). As depicted in Figure 5.4, requirements flow down commences when new or updated regulatory requirements are identified by one of the organizational elements that is a regulatory responsibility area (i.e., Environmental, Transportation and Health and Safety in our examples). The new or updated regulatory requirement then enters into the stakeholder review process where applicability to each organizational element is evaluated and decisions are made on implementation strategies. Membership of the stakeholder review process includes representatives from each organizational element who are then responsible for communicating the new or updated requirements to their respective organizational element.

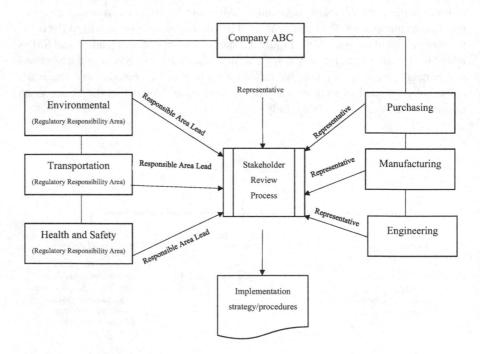

FIGURE 5.4 Stakeholder review process.

The stakeholder review process is certainly not limited to environmental requirements and can be utilized for all requirements and protocols proposed for implementation within the company or corporation. Many companies have a similar process for reviewing and approving requirements already built into their organization, making the integration of environmental regulatory requirements relatively straightforward. The stakeholder concept has applicability to companies of all sizes, with the only real variables being the volume and complexities of applicable regulations and the number of stakeholders participating in the process. To reiterate, the important aspect of the concept is that there is a defined process that ensures a consistent approach and inclusion of all relevant stakeholders in the decision process.

5.2.1 CORPORATE-LEVEL PROCESSES

At the corporate level of an organization, the architecture of the requirements flow down process is commonly defined. The corporate-level processes include defining the policies and rules on how the flow down process will be conducted and the associated roles and responsibilities. These corporate-level policies should include detailed descriptions on how the requirements flow down process will function, including clear definitions of the following:

- Regulatory responsibilities areas within the organization
- Roles and responsibilities of regulatory responsibility areas
- Stakeholder review process membership
- Authorities for decision-making within the stakeholder review process
- How stakeholder reviews, decisions and implementation strategies are documented
- Requirements tracking

It is important that this process be well defined and openly communicated throughout the organization so that it is clearly understood how requirements and requirement flow down are governed. Establishing a clearly defined requirements process builds efficiency into the organization by providing a consistent process for identifying requirements, determining applicability and developing implementation strategies. The efficiencies that can be realized from a well-defined requirement flow down process can be greatly enhanced if the process is integrated into or parallels existing corporate processes. In other words, if the process for governing environmental requirements can be integrated into an already existing corporate process, there will be an inherent level of familiarity to personnel. This approach is preferred to creating a whole new and separate process for environmental requirements that would require an increased expenditure of time and effort for development, documentation and training. The ability of a company to integrate an environmental requirement flow down process into the organization is of course highly dependent on a multitude of variables. Smaller companies or companies with minimal regulatory requirements are most likely faced with an easier challenge of governing environmental requirements. Larger companies operating in highly regulated industries have increased compliance obligations and a greater need for a well-defined and integrated requirement flow down process. In either case, it is important that there is a consistent approach for environmental requirements flow down into each relevant level of the company.

5.2.2 ORGANIZATION-LEVEL PROCESSES

Activities that should be established at the organizational level to support requirements flow down need to include identifying personnel to fulfill the roles defined at the corporate level and establishing organization-level protocol or procedures that interconnect and complement the corporate-level process. For organizations that are identified as regulatory responsibility areas, the following actions should take place or at least be strongly considered, depending on the size of the company and available resources:

- Assign a regulatory responsibility area lead
- Define organization-level roles and responsibilities for regulatory responsibility area lead
- Assign regulatory SMEs (Note: This may also occur at the work unit level.)
- Define roles and responsibilities for SMEs
- Identify training and qualification requirements for area lead and SMEs
- Identify procedures and instructions needed to integrate with corporate process

    For organizations within the company that are not identified as a regulatory responsibility area, the following should be established to support the requirements flow down process:

- Identify representatives to participate in the stakeholder review process
- Define how decisions resulting from stakeholder reviews will be communicated internally
- Define processes for incorporating requirements and implementation strategies approved by the stakeholder review process

Depending on the size of the company, there may not be a need to go beyond the organizational level of the company for establishing processes and identifying personnel to support the requirements flow down and stakeholder review processes. What really dictates the level of organizational participation is how adequately identified requirements and implementation strategies are being communicated to the relevant parts of the organization. Another important aspect that needs to be considered is the adequacy and appropriateness of the implementation strategies. If the people responsible for implementing compliance strategies are not involved in the decision-making process or are not afforded a means to provide input, then there is a high probability that the implementation strategies may not be effective. While ineffective implementation strategies may not always result in a regulatory noncompliance, they certainly can if they are not corrected or revised to adequately apply to the intended work process. Revising an inadequate implementation strategy requires identifying why it is inadequate, making the appropriate changes and going through the stakeholder review process again. This would obviously require additional time and resources and can be avoided the majority of the time if the people performing the work have input into the development of the implementation strategies affecting their work activities.

5.2.3 WORK UNIT–LEVEL PROCESSES

Integrating work unit–level organizational aspects into the requirements flow down process can be very beneficial to companies that have a large number of regulated work activities or processes across multiple organizational elements. To better understand what a work unit level might look like within an organization, Figure 5.5 provides an expanded version of the example organization chart for Company ABC. As depicted in the figure, work unit levels within an organizational structure are typically defined by a definitive scope of work and associated roles and responsibilities. For instance, Company ABC's environmental organization consists of a Water Monitoring Group responsible for wastewater and storm water and an Air and Waste Group responsible for air regulations and hazardous waste.

The Health and Safety Group has a similar internal structure with an Industrial Hygiene Group and an Industrial Safety Group. The point to be made here is not

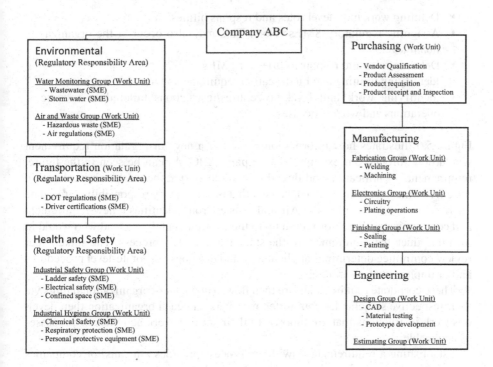

FIGURE 5.5 Organizational work unit levels.

necessarily how to divide responsibilities within an organization but to illustrate that each of the groups at the work unit level has defined areas of responsibility and needs to be integrated into the requirements flow down process. The importance of work units participating in the requirements flow down process should not be limited to organizational elements that are regulatory responsibility areas. Other organizational elements may have activities or processes that require implementation of approved strategies to ensure compliance with environmental regulations and may also need to be integrated into the overall flow down process at the work unit level. In Figure 5.5, the organizations that have been identified as regulatory responsibility areas in previous figures are on the left of the organization chart. As previously stated, these areas are responsible for identifying and tracking applicable regulatory requirements within their respective areas of responsibility and communicating this information to the stakeholder review process. On the right side of the organization chart are examples of organizational elements with work units that are responsible for implementing regulatory requirements that apply to their activities and processes. As previously discussed, all of the organizational elements participate in the stakeholder review process and are responsible for developing protocols and identifying roles and responsibilities within their organization that integrate into and support the corporate-level requirements flow down process. For companies that include integration at the work unit level, this will require:

- Defining work unit–level roles and responsibilities
- Assigning regulatory SMEs (Note: This may also occur at the organizational level.)
- Defining roles and responsibilities for SMEs
- Identifying training and qualification requirements for SMEs
- Identifying work unit–level procedures/instructions tailored to specific operations and work processes

Figure 5.6 illustrates how information regarding a new air regulation requirement would flow through our example of Company ABC. As can be seen in the figure, requirements identification and flow down would start at the work unit level within the environmental organizational element that is a regulatory responsibility area. The air regulations SME within the Air and Waste Group identifies a new requirement and communicates that information up to the environmental area lead who would in turn introduce the requirement to the stakeholder review process. The stakeholder review committee determines applicability and develops corporate-level procedures and/or implementation strategies.

These corporate requirements are then flowed down to all organizational units via their respective stakeholder committee members and each organization then develops work instructions that are tailored to their specific work activities or processes as applicable.

Establishing a requirement flow down process provides a means for companies to consistently and efficiently communicate regulatory requirements to all relevant elements of the organization. A company's ability to consistently progress from the

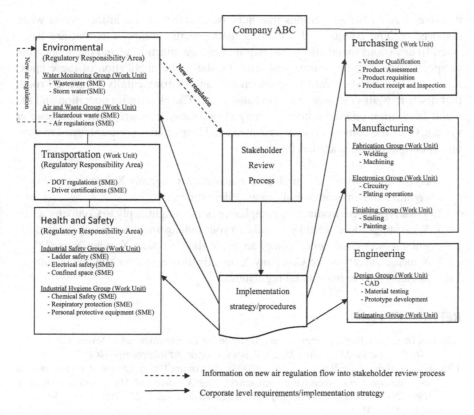

FIGURE 5.6 New requirement information flow.

identification of regulatory requirements to the implementation of effective compliance strategies is an important element of a sustainable environmental compliance program. In order to ensure consistency in requirements flow down, the process including roles, responsibilities, information flow and authorities must be documented and readily accessible to all organizational elements.

5.3 APPLIED LEARNING

Review the case study and respond to the questions that follow.

5.3.1 CASE STUDY

Company X designs and builds various components for the aircraft industry. The company has 200 employees and all of its operations are conducted in a single facility. Company X purchases its raw materials and supplies from multiple vendors and has them shipped directly to the factory. Some of the components manufactured by Company X include electronics that are completely designed and assembled within the company's factory. The manufacture of the electronic components requires use of solvents and plating solutions that must be managed as hazardous waste when disposed of. The company also manufactures specific components that are fabricated from various metal alloys that require welding, machining and the application of special epoxy resins, paints and other finishes. The application of these epoxies, paints and finishes must be conducted in an environmentally controlled room that discharges air emissions up an exhaust stack. The electronics operation and the metal fabrication operation both discharge rinse water into a regulated wastewater system. Once the components are assembled and inspected, they are shipped to various aircraft companies around the world.

1. As the environmental compliance manager for Company X, how might you organize the environmental group or department?
2. What type of environmental compliance issues might apply to Company X?
3. What suggestions would you make to your management regarding the integration of environmental compliance into the Company X organization?
4. Which elements of the Company X organization might be affected by new or changing environmental regulations?

REFERENCES

Carlile, Liz. (2011). Report. International Institute for Environment and Development, 2011. *JSTOR*. Accessed February 25, 2020. www.jstor.org/stable/resrep01461.
Thompson, Leigh, & Brajkovich, Leo F. (2003). Improving the creativity of organizational work groups [and executive commentary]. *The Academy of Management Executive (1993-2005), 17*(1), 96–111. *JSTOR*. Accessed February 25, 2020. www.jstor.org/stable/4165931.

6 Requirements and Implementation Documents

Chapter 5 included discussion on flowing down requirements within the organization and establishing a centralized stakeholder review process to act as a clearing house and governing body for regulatory requirements. To recap, the stakeholder process was a centralized committee or group where newly identified regulatory requirements were presented, applicability determined and implementation strategies and or procedures developed and subsequently communicated throughout the organization. This requirement flow down process involves the development of procedures and/or policies at the corporate, organizational and work unit–levels of the organization that either govern or support the requirements flow down process. In this chapter, we will discuss a conceptual hierarchy of these documents and provide discussion and examples of how the procedures and/or instructions at the various organization levels can be integrated into a requirements governance and flow down process.

6.1 CORPORATE-LEVEL PROCEDURES AND IMPLEMENTING DOCUMENTS

Corporate-level procedures and policies typically represent a collection of requirements or policies that govern the various activities and functions of the company. These corporate- or company-level policies and procedures represent the top level of the document hierarchy within the company and should establish the upper-level requirements and protocol that describe the bounding conditions organizational elements within the company operate to. As an example, corporate- or top-level policies can be established for the various primary functional aspects of a company such as human resources, finance and budgeting, records management, safety and environmental compliance and, of course, a governance process for requirements management and flow down.

Table 6.1 provides an example of the structure and content of corporate-level policies for the business functions.

The level of details in corporate- or company-level procedures is obviously dependent on the size and complexity of the overall organization. Smaller companies focused on a single product or service with fewer organizational components are more likely to have the ability to operate under a set of policies and procedures at the corporate or company level. Larger and more diversified companies that provide

TABLE 6.1

Company ABC Corporate-Level Policies and Associated Content

Business Element	Policy Content
Corporate governance	1. Describes process for managing requirements within the company 2. Defines roles and responsibilities for organization elements regarding the requirements management process 3. Authorities for decision-making are defined 4. Establishes stakeholder review process and organizational participation requirements 5. Defines requirements tracking system/process 6. Identifies documentation requirements
Human resources	1. Personnel policies defined 2. Preemployment requirements identified 3. Affirmative action requirements identified 4. Leave policies defined
Finance and budget	1. Financial system defined 2. Financial reporting requirements defined 3. Cost accounting and allowable costs defined 4. Contracting and procurement requirements defined
Records management	1. Records identification requirements defined 2. Records retention requirements defined
Safety management	1. Corporate safety philosophy defined 2. Safety requirements for organizational elements defined 3. Corporate injury/illness reporting requirements defined
Environmental compliance	1. Corporate environmental policy defined 2. Responsibilities for managing environmental regulations defined 3. Responsibilities for regulatory compliance defined 4. Corporate environmental reporting requirements defined

a multitude of products and/or services and operate under a more complex organizational structure typically benefit from corporate-level policies and requirements that require lower tiered organizational components to develop procedures and processes to comply with corporate-level requirements and obligations. Whether a company is large or small, it is important that corporate-level policies and procedures provide for clear and unambiguous actions that organization elements must undertake to support and meet corporate policies and requirements.

6.2 ORGANIZATION-LEVEL PROCEDURES AND IMPLEMENTING DOCUMENTS

Organization-level procedures are typically a collection of documents that are developed to describe the processes and actions that a specific organizational element (i.e., environmental compliance department) will implement to meet the corporate- or company-level requirements. The concept here is to create an integrated system that allows each organization to develop implementing documents that are tailored to their specific activities and operations. From an environmental compliance standpoint, this approach allows organizational elements to create procedures or instructions that will ideally maximize operational efficiencies to the extent possible while integrating the necessary steps for regulatory compliance. Integrating specific steps for environmental compliance into operational procedures or implementing documents is often less disruptive to work flow and provides for a consistent approach to compliance. Using the information in Table 6.1, we can see that at the corporate level, the requirements for the environmental compliance organization have been defined to include responsibilities for managing regulations, environmental compliance and environmental reporting. Per the model discussed here, the environmental compliance organization would develop implementing documents that would provide the procedural steps for how the organization will actually manage environmental regulations, compliance and reporting. As an example, the environmental compliance organization would develop a procedure that identifies the roles, responsibilities and process steps for identifying regulatory requirements, assigning regulatory SMEs, determining applicability, reporting regulatory requirements to the stakeholder review process and requirements tracking. A separate procedure could be developed for environmental compliance that would describe the roles, responsibilities and processes for managing environmental compliance within the company. This procedure could include steps for assessing operations and activities within the company, identifying nonconformances, reporting requirements, corrective actions and tracking/trending compliance issues. In both cases, these could be a single procedure or a series of procedures depending on the size of the organization and the complexity of the regulatory setting within the company. Figure 6.1 illustrates the relationship between corporate-level requirements and the environmental compliance organization's implementing documents.

As depicted in the figure, the corporate-level requirements define what must be done and the organization-level implementing documents describe the specifics of how it is going to be done and who is going to do it!

It is important when developing strategies for sustainable environmental performance that the use of implementing documents enables operational efficiencies as well as compliance. This is critically important because procedures and other implementing documents that become too voluminous or unnecessarily complex lose their usability and effectiveness. A strategy that incorporates a defined process for controlling the development, review, revision and collateral effects of procedures will help in mitigating this potential concern.

6.3   PROCEDURAL GOVERNANCE

A systematic approach for managing procedures is an essential element for any compliance strategy. Without a defined process for governing how procedures are developed, reviewed, approved, maintained and revised, the interorganizational connectivity becomes difficult, if not impossible, to maintain and companies can

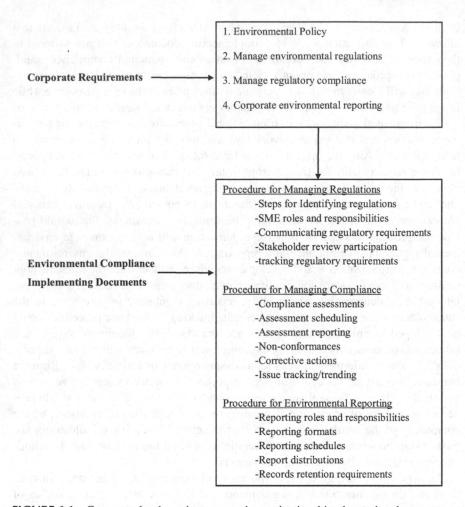

FIGURE 6.1 Corporate-level requirements and organizational implementing documents.

rapidly lose control over implementation of requirements and regulatory compliance. Considering the dynamic nature of environmental regulations, organizations can benefit from establishing a system for procedural governance that invokes participation and ensures consistency throughout the organization as a whole.

6.3.1 DEVELOPMENT AND APPROVAL

When developing procedures or requirement documents within a company or an organization, it is beneficial if there is an established document template or format. Established formats help ensure that procedural content is addressed consistently throughout the company. An example procedure template is provided in Figure 6.2.

The sectional content included in a procedure template can, of course, be adjusted to best accommodate company needs. This is especially true for the procedural steps

Procedure XYZ   Effective Date xx/xx/xx   Expiration Date xx/xx/xx   Rev. 0
1.0   **Purpose** – Defines the purpose of the procedure. Why it is needed and what is to be accomplished.
2.0   **Scope** – Defines the activities, operations or processes that are covered by the procedure
3.0   **Exceptions** – Identifies any exceptions to the procedure
4.0   **Prerequisites** – Defines any required prerequisite actions that must be taken prior to initiating procedure
5.0   **Definitions** – Definitions of key words and technical terms
6.0   **Acronyms** – Listing of acronyms used in the procedure
7.0   **Procedural Steps** – Description of actions and steps to be taken in a specified order to accomplish purpose of procedure
8.0   **Records** – List of records generated by implementing procedure
9.0   **Training** – Listing of required training
10.0   **References** – Listing of referenced procedures and document
11.0   **Revision History** – Revision history of the procedure and summary of associated changes
Page 1 of n

FIGURE 6.2   Procedure template example.

in Section 6.0 of the example, which can be broken down into further subsections to capture the desired level of details and instructions.

Company protocols for procedure development and the associated approval process are commonly itself documented in a procedure. This may seem a bit obtuse at first glance, but the proverbial 'procedure on procedures' is one of the primary building blocks for developing a controlled and consistent process for managing requirements and flow down through implementing documents such as procedures and work instructions. A procedure for the development and approval of procedures is necessary to not only promote consistency in content as previously discussed, but also to ensure that procedures have defined levels of approval. Approval of procedures most often involves multiple signatures that can vary based on the organizational level of the procedure and the management structure of the organization. A typical approval chain for a procedure or other implementing document will include the author, the work supervisor responsible for personnel implementing the procedure, the work supervisor's manager and in some instances additional levels of management. Depending on the scope of the procedure and the organizational structure of the company, an approval chain for a procedure may include facility management personnel, safety personnel, compliance personnel or representatives from corporate legal. Since approval chains can vary based on the organizational level and scope of the procedure, a single predefined approval chain is probably not going to work for larger companies with multitiered organizational structures and multiple internal stakeholders. Determining the right approval chain is intuitively part of the procedure development and review process and should be integrally associated with identifying who has responsibilities or a stake in the operations or processes within the scope of the procedure.

6.3.2 STAKEHOLDER CONSIDERATIONS

Including the right people in the development and review of a procedure can make a significant difference in the quality of the procedure. By properly identifying all of the organizational elements that can be affected by the procedure and providing a means for their input, unforeseen impacts of implementing the procedure can be minimized. Chapter 5 presented discussion on the stakeholder review process relative to the process of managing requirements and requirements flow down. This concept can apply to any level of procedure development within an organization in that it is always beneficial to include the appropriate stakeholders in the procedure development, review and approval processes. This is not meant to imply that every level of the organization has to have a formal stakeholder review process equivalent to that presented in Chapter 5. However, stakeholder input should always be considered when developing or revising procedures, including those regarding regulatory compliance. There are numerous means by which stakeholders can provide input. Stakeholders can participate in the development process upfront, be involved in the review of drafts of procedures or procedure changes, be involved in the approval chain or be involved in all of these phases. It is also important to consider that there will be varying levels of interest among the stakeholders based on the level of impact or involvement in the activities that are to be covered in the procedure. These varying interest levels may dictate how involved in the process a specific organizational element wishes to be. Some groups within the company may want to be involved in every stage of the process and some may wish to only be involved in the review. There is no hard and fast rule that can be applied to every company or corporation. The important message here is that some flexibility needs to be built into processes for stakeholder participation. Experience has shown that stakeholder inclusion upfront can reduce the potential for numerous and ongoing procedure revisions that become necessitated by incomplete or inadequate procedure content.

Identifying the correct stakeholders for a given procedure can be accomplished by using several different approaches. One approach would be to provide each organizational element within the company the opportunity to participate in the development and review of all procedures. This approach has some benefit to both larger companies with more complex organizational structures and smaller companies with simpler structures. This approach ensures at a minimum that every organizational element has the chance to review the procedure and determine if there will be an impact on their operations or activities. A second approach is to define stakeholder groups for each organizational element. As an example, there could be a defined stakeholder group that would participate in the development and review of any procedure coming out of the environmental department. Lastly, there is an approach that involves determining stakeholders for each individual procedure. There are undoubtedly pros and cons to each of these methods that will vary based on individual situations. Some general pros and cons for each approach are presented in Table 6.2 for consideration purposes.

6.3.3 CONTROLLING CHANGES

Once approved for use, the majority of procedures, work instructions and other implementing documents will invariably require some level of revision during the time frame that they are actively in use. There are numerous factors that come into play that normally require revisions to procedures. Many times changes are identified during the initial implementation of the procedure that are related to the content within the procedure that doesn't reflect how the actual operations are performed. In other instances, procedural revisions are needed to add steps to adequately address operational safety concerns or regulatory requirements. Other factors that often

TABLE 6.2

Comparison of Approaches for Stakeholder Inclusion

Method	Pro	Con
Each organizational element participates	1. No inadvertent exclusions 2. Full company participation and awareness	1. Every organizational element must review every procedure 2. Longer development and review time frames
Defined stakeholders for each organizational element	1. Consistency in reviews 2. Shorter development and review time frames	1. Risk of procedure impacting organizational element not represented in defined stakeholder group
Stakeholders determined for each procedure	1. Limits the number of organizational elements reviewing each procedure 2. Shorter development and review time frames	1. Lacks consistency in reviews 2. Stakeholders may need to be reevaluated with major procedure revisions

necessitate procedure provisions are changes to operations, evolving safety requirements and changes in regulatory requirements. With such a large percentage of procedures potentially going through revision at some point, it is important that procedural
changes are controlled to ensure that only the current approved version is being used.
Without a process for procedure change control, companies run the risk of losing version control of procedures, which can result in numerous and outdated versions of a
procedure being implemented within the company. It goes without saying that implementing outdated procedures can present increased vulnerabilities to companies in
regard to worker safety and health, operational efficiencies and regulatory compliance. A fundamentally sound procedural control program does not have to be overly
complex and can be effective by incorporating the following elements at a minimum:

1. Procedure title and number
2. Page numbers
3. Approval date included on procedure
4. Effective date included on procedure
5. Expiration date included on procedure
6. Revision number included on procedure
7. Method for notifying personnel of procedure revisions
8. Database or another system where employees can access and verify current
   versions

It is advantageous to include the procedure title, number, approval and effective
dates as well as the procedure revision number on each page to preclude any confusion or misinterpretation if the pages become separated during implementation. This
may seem like a formality, but there have been more than a couple of instances of
procedures being used for operations that actually contained pages from older revisions that had gotten commingled in the current revision of the procedure. Having
the revision number and the aforementioned dates in the header or footer of every
page provides a simple means to assist in preventing this type of occurrence. These
simple elements coupled with a consistent method to notify personnel of procedure
revisions and a document control system where employees can readily access current
revisions will provide the necessary components for controlling changes to procedures and other documents.

6.3.4 CONFIGURATION MANAGEMENT

Not only is it important to control procedural changes to ensure only the most current and correct procedures are being used, but it is also important to understand
that other procedures or organizational aspects may be affected by changes to a procedure. As an example, oftentimes environmental compliance procedures include
requirements that multiple organizational elements within the company must comply
with. As discussed in Chapter 5, ideally these environmental requirements would
be flowed down to the organizational elements where they would be incorporated
into their organizational procedures. If the environmental compliance procedure is

revised, then every procedure or document within the company that has incorporated requirements from the environmental compliance procedure must, at a minimum, be evaluated to determine if similar changes are necessary. Without this type of configuration management, it is fairly easy to understand how quickly procedures and other implementing documents could become disassociated and out of sync with current company or regulatory requirements. To be complete, a configuration management program needs to also consider other operational aspects that could be impacted by procedure changes:

1. Training associated with or required by procedure
2. Prerequisite actions identified in procedure
3. Associated procedures/companion procedures
4. Forms
5. Checklists
6. Diagrams
7. Regulatory permits

The complexity of configuration management will obviously increase with the complexity of the organization and the number of procedures and documents being utilized. Instituting an effective configuration management program can be directly associated with the procedure development process and the required content for procedures. Referring back to Figure 6.2, if the template includes required content such as prerequisite actions, required training and referenced procedures and documents, then these aspects all become considerations for configuration management if and when there are changes to the procedure. Configuration management is also where stakeholder participation can play an important role. When stakeholders are involved in the development and review of procedures, the ability to identify potential impacts of procedure changes is greatly enhanced.

6.3.5 PERIODIC REVIEWS AND UPDATES

Procedures and implementing documents should undergo periodic reviews to ensure that they are still adequate and information within the procedure is still accurate and relevant. Establishing review and update schedules is an efficient and proactive means for maintaining procedures at all levels of the organization. Procedure review cycles are typically identified in the company document that governs the development, review and approval of procedures (the 'procedure on procedures'). Without an established schedule for reviewing and updating procedures and other critical documents, a company can create a reactive environment where it can become increasingly difficult to keep up with and track all the needed changes. Quality assurance standards such as ISO 9001, NQA-1 and others recommend that procedures and other implementing documents undergo review and update every three years at a minimum. Many approaches used in industry implement review and update cycles that require an annual review of procedures and implementing documents in conjunction with updates every three years. To further clarify this approach, the annual review requirement is intended to be an evaluation of the procedure to determine if the procedure is still accurate. Annual reviews are commonly a review by the procedure owner or responsible person and are often used to capture administrative-type changes (e.g., contacts, phone numbers), typographical errors, references, etc. The three-year review and update is intended to be a complete and thorough review of the procedure that includes participation by all the stakeholders. Three-year reviews typically involve a much more detailed evaluation of the document and often result in substantial changes. At the completion of a three-year review and update, a procedure is usually issued a new revision number and new approval, effective and expiration dates.

6.4   ACCESSIBILITY OF PROCEDURES

Having a set of procedures that are accurate, up-to-date, properly controlled and under robust configuration management will be of minimal value if they are not accessible within the organization. The accessibility of procedures is paramount for every aspect regarding procedure and document management. Ready accessibility is commonly made possible through the utilization of electronic storage and management of documents using databases or other document management software. Obviously, procedures need to be available to those responsible for implementing the procedures within their organizations and areas of responsibility. As previously discussed in Section 6.3.3, having procedures readily accessible allows for users to verify that they are using the most current and correct version of the procedure. With the technology today, there is ample opportunity to utilize handheld electronic devices to access electronically stored procedures in the field, which can add a great deal of efficiency to procedure use and control. However, there are many operations in the field where handheld devices are not practical or allowed and hard copies of procedures are needed during operations. Ready access to current procedures allows for this simple verification to take place prior to the commencement of operations. Procedure accessibility is not only important to the end user, it is also critical to internal stakeholders, procedure owners, administrative personnel, managers, training personnel, safety personnel, etc. Ideally the entire organization should have access to all procedures. There are of course exceptions to this when procedures involve sensitive, proprietary or classified information. In these cases, access obviously has to be restricted to the appropriate audiences, but however limited the audience, the information still has to be accessible for all of the same reasons mentioned above.

6.5   APPLIED LEARNING

1. Discuss the difference between corporate- and organizational-level procedures.
2. Discuss why having a 'procedure for procedures' is important.
3. What are the important elements that should be included in a 'procedure for procedures'?
4. Discuss the consequences that can result from the inadequate management of procedures and implementing documents.

7 Critical Implementation Elements and Attributes

7.1 ENVIRONMENTAL JUSTICE

The concept of environmental justice came to the forefront and took form in the 1980s, resulting from the pollutants companies released into the environment across the country. These pollutant events were discovered primarily in low-income and disadvantaged areas where people lived and worked. Environmental justice is defined by the Environmental Protection Agency (EPA) as 'the fair treatment and meaningful involvement of all people regardless of race, color, national origin, or income, with respect to the development, implementation, and enforcement of environmental laws, regulations, and policies.' The goal of the EPA is that all people are provided the same degree of protection from environmental health hazards, affording them a healthy environment where they live and work (https://www.epa.gov/environmentaljustice).

The EPA in 1992 established the infrastructure for environmental justice through the establishment of the Office of Environmental Justice (OEJ). This office has been actively engaged in coordination of the EPA's efforts to address environmental justice issues across the United States (https://www.epa.gov/fedfac/environmental-just ice-fact-sheet). A significant part of ensuring environmental justice is the enactment of the National Environmental Policy Act (NEPA), which was enacted in 1970, more than two decades before the establishment of the infrastructure to support the office of environmental justice. This act requires entities to conduct a thorough impact of the proposed business and its impact on the environment and people residing within the area.

Studies have shown that environmental hazards are not distributed equally across the United States. It is recognized that poor people and poor people of color are exposed to more pollution than rich people and Caucasians. Environmental justice involves democratic decision-making and community empowerment (Cole and Foster 2001). Poor people and people of color in the past have been underrepresented in the democratic decision-making process when determining where to locate different types of factories and industries that typically would emit hazardous substances to the environment. Beginning in the mid-1980s was a recognition that people of color in urban and rural areas might be exposed to greater levels of environmental risks than other populations. Large groups of minorities compared to other groups tend to live closer to facilities that handle, manage and dispose of hazardous waste and chemicals. In previous years, minority populations were not actively involved in the decision-making process. To remedy the issue of locating hazardous operations completely in areas where people of color and disadvantaged populations have their residence, the EPA requires that a company must conduct a risk assessment to evaluate demographic and environmental data (Rhodes 2003).

The concept of environmental justice is not always adequately taken into consideration when making decisions on placement of high-risk industries that can emit hazardous substances and the impact on the neighboring communities. However, because of community activism, there are groups that are vigilant of the types of industries that are attempting to relocate to their environment and the impact they may pose on the people and the environment. These groups in some ways are more powerful in ensuring protection of the environment and life itself.

It is these aspects that create an undeniable connection between environmental justice and corporate responsibility. Corporations bear the responsibility to their workers, stakeholders and surrounding communities to ensure that the potential for exposure to toxins is minimized and that they are transparent in any information related to product toxicity, pollutant discharges and exposure potential. Corporations need to be responsive to environmental justice concerns and should be held accountable in their decision-making processes relative to locating operations involving toxic operations and discharges of pollutants. In addition, corporations as well as regulatory agencies need to ensure that complete and accurate information is made available to communities and interested parties and that they are allowed to actively participate in the decision-making processes. The historical trends of environmental discrimination against economically disadvantaged and minority communities have been associated with industry withholding or denying access to information regarding industrial toxins and associated risks (Markowitz and Rosner 2002). The remedy for appropriate and equitable environmental justice cannot be placed solely on regulatory agencies. Corporations need to embrace a culture that not only acknowledges the need for environmental justice and transparency, but emphasizes them as critical components of corporate social responsibility.

7.2   TRAINING AND KNOWLEDGE RETENTION

Companies struggle constantly with how much financial resources is necessary for training and knowledge retention as cost of training continues to rise and workers are required to become knowledgeable of various disciplines or topics and sustain some type of certification to demonstrate proficiency and credibility in the chosen field. This is especially true for technical workers, including engineers and scientists. Securing and retaining certifications can be a financial strain on an individual and a company, depending on the number of certifications an individual is required to possess and the number of workers that are holding certifications. An effective training strategy that provides a road map or clues to management to answer critical questions that can help them determine what training is important to propel organization success is an important attribute of an overall business strategy. The following questions at a minimum should be explored:

- Who to train?
- When to train?

- What type of training is required?
- What training should be conducted internally?
- What training should be obtained from external vendors such as a conference or training institution?
- How much to set aside annually for training?
- What does the refresh frequency look like for the various types of training required of workers?
- How to retain train and qualified workers?

Answering these questions and leveraging the answers to inform decisions do not always occur. Or if they occur, they are not comprehensive enough to provide the level of information needed to accurately diagnose and remedy shortfalls in training. Too often leaders fail to focus on training and knowledge retention within their organizations. Some assume that because they hire highly skilled and qualified workers, they will have access to the skills needed to achieve the goals of the organization. However, if the skills needed to achieve goals are to be available when needed, attention must be paid to knowledge retention. The traditional method used to impart knowledge for purpose of learning is typically performed in a traditional classroom setting. This style of knowledge impartment for the purpose of learning may not be the most effective method for retention for some professionals. Therefore, it is necessary to utilize various available techniques to increase retention. Some of these methods may include the use of techniques such as brief storytelling, quizzes, review and summarization exercises, demonstration of concepts, active participants, applying the concept to real-life activities, hands-on learning and bite-size learning.

There are three methods primarily used by people to learn: auditory, visual and a combination of both. Although an individual may prefer and is successful in absorbing information using one or more of the learning techniques, when applied or active learning is added to the learning process, an individual has the opportunity for better retention and application of the material exposed to in a traditional classroom setting. There are many aspects of training that have been deployed to prepare students and professionals to be productive subject matter experts. However, there are two processes that are very effective in helping an individual retain the information and put it into use: The applied and active learning processes are extremely important in the development of technical workers such as engineers and technical professionals to improve their critical and strategic thinking capability.

It has been stated by many scholars of psychology and education that critical thinking can be taught in the classroom. Also, many teaching organizations have designed and taught classes on critical and strategic thinking to workers in their workplace. As to how effective these courses are in achieving their objective, one could surmise that it depends on the individual, presentation of the course materials and the type of reinforcement provided by the instructor. Critical thinking when taught in the classroom requires the engagement of different teaching techniques that incorporate active and applied learning. Some professionals find it easier to learn and retain the information when the process of active or applied learning is deployed effectively.

The process of active learning involves having the student engaged in the process of learning through reading, writing, discussions and solving problems, both complex and simple. The process of active learning can be used to promote critical thinking. Some active learning activities that can be used to facilitate critical thinking in the classroom include (Kim et al., 2013):

- Use scenarios that will require students to apply their knowledge by providing a justification for their responses
- Introduce problem-solving activities that require students to engage in data analysis, evaluation and synthesis
- Utilize problems that have more than one possible answer, requiring students to consider different alternatives and views
- Use open problems that require students to apply key technical concepts such as engineering, biology, chemistry and physics to the solving of problems encountered every day
- Utilize open problem-solving in small groups and encourage participation and input from all group members
- Use collaborative learning technique to group projects that require the use of dialogue and social interaction in learning and carry out projects that require the capabilities and abilities of others to reach a decision or complete a task
- Write essay and research papers

Applied learning is an approach that has been utilized through partnership and collaboration with corporations and educators and is also used by corporations to enhance knowledge and skills of workers. This type of learning is an approach that actively engages students and professionals in directly applying the skills, theories and models that they have learned mostly through classroom learning. Applied learning coupled with traditional classroom learning provides an individual the optimal opportunity to learn and retain critical information. In addition, it can spark individual creativity because applied learning provides reinforcement to classroom learning presenting the ability to see how theories and methods are applied to solve real-life problems. Many college institutions require students to be involved in at least one internship in order to successfully complete their degree requirements. Internship is a good example of applied learning and is a great way to gain experience in theories and concepts learned in a classroom. Some of the benefits of applied learning can include: an effective tool to convert classroom knowledge to real-life application and a great way to network with others with similar specialties and similar knowledge, gain professional experience and skills that are being sought after by employers, assist the learner in developing a better understanding of course materials they were exposed to. Applied learning methods include:

- Internship – the student or person in training works with a company to gain experience in their area of study. The experience gained can be transferred into full-time employment

- On-the-job training
- Shadowing
- Volunteer service

Other methods that can be used to assist in training and knowledge retention include the use of team-based learning (TBL) and microlearning. TBL is a type of small-group learning that takes advantage of the knowledge and social skills of all members of the group as they interact, collaborate and share information. It is important to recognize that when professionals are working together in a team environment, learning is taking place and each team member should exit the team with more knowledge than prior to functioning as a member of that team. This type of learning occurs when people are afforded the ability to work on a team of workers having similar or different skills that are used to contribute to completion of a project or a task. This type of learning is extremely effective because it provides exposure to various topics as well as the ability for team members to learn from each other through discussion that is more likely to be repetitious in nature.

According to Sibley and Ostafichuk, team-based learning generally consists of five to seven participants, although it can be larger to facilitate problem-solving. The essential elements of TBL are: (1) the group must be appropriately formed as well as managed, (2) everyone must be accountable for their individual work and the deliverables of the group, (3) group assignments should be designed to promote learning and team development, and (4) students must be provided timely feedback on performance. These elements are similar elements needed for team cohesion and effectiveness. Let's consider how the four elements of TBL are applied to team learning. When forming a team, great consideration is given to the propose of the team, the work that is expected to be accomplished and the skill sets needed by team members to render the project a success; the team must be appropriately managed and held accountable for their individual tasks as well as supporting other members of the team; timely feedback of performance is provided to team members by members of the team as well as the team manager; group assignment promotes learning through the interactions between the professionals on the team with diverse skills and knowledge, which is shared through interactions. TBL can only serve its complete purpose and become a competitive advantage when the team functions in an inclusive environment, every member is actively engaged and assumes accountability and responsibility for their individual tasks and actions.

The philosophy of microlearning is gaining traction and popularity as many companies and educators gain knowledge of the concept and the benefit it brings to a learning environment. The concept of microlearning sometimes called bite-size learning is a process where small bursts of information are introduced to increase retention. Poor knowledge retention occurs often primarily because of having too much information to absorb and retain at a time. Microlearning is like feeding small amounts of food frequently so that it can be digested and keep the metabolism fueled. Oftentimes in a learning environment or a work environment, people may have difficulty retaining information and instructions due to the amount, diversity and complexity of information presented. This inability to retain information in the workplace can lead to conducting a task improperly or injury to personnel when safety practices are forgotten or implemented incorrectly. Many Fortune 500 companies use mentoring and coaching as an integral part of their employee development and retention strategy. These processes have been successfully used by many organizations to build and strengthen the skills of professionals and are viewed as low-cost measures that can yield maximum benefits when implemented properly.

Mentoring is an activity that makes use of pairing less experienced and knowledgeable professionals with more knowledgeable and experienced professionals. A mentoring relationship is rewarding for both parties from a personal and professional standpoint. Mentoring provides an opportunity for participants to develop communication and professional skills, career advancement and to discover different approaches to solving problems and approaching issues as they arise. The mentoring process focuses on the future and, therefore, it is a tool that leaders use mostly to prepare critical staffing for future assignments and roles. Some potential benefits of having an effective and active mentoring program include enhanced employee engagement, increased job satisfaction among employees, increased employee retention and enhanced knowledge and skills.

Coaching is a personal development activity where an experienced and knowledgeable individual supports another who is less experienced or knowledgeable in the discipline or skills providing guidance and training opportunities. The process of coaching generally focuses on enhancing knowledge and skills for the present. The use of coaching has a negative stigma for some as it has been used to help correct performance issues when professionals fail to meet the expectation of the roles and responsibilities within their organization.

Training is key to implementing a comprehensive, effective environmental management system within an organization. A comprehensive training program not only has provisions for initial knowledge gaining, but also takes into consideration the necessary repeat or refresh frequency of training material and knowledge. For example, training refreshment and material updating will most likely be needed when there is a change in regulatory requirement big or small. Training refresh must be tailored to the changes and should not be seen as an opportunity to provide initial training to workers unless there has been an assessment to determine the need to provide initial training as well as the changes in regulatory requirements. Any or all of the training methods discussed in this section can be used to enhance workers' knowledge of environmental laws and practices within their organization.

7.2.1 APPROPRIATE KNOWLEDGE FOR IMPLEMENTATION

An organization is unable to function effectively without qualified professionals with the level of knowledge to think creatively, devise solutions and prevent issues from occurring. The development and retention of practitioners that are key to technology development, engineering design, research and development and other roles that are necessary to provide products and services to a global economy is challenging, especially in today's climate with an above average economy presenting a plethora of technical opportunities available for skilled candidates. Training must be a part of the strategy to ensure practitioners are equipped to handle the jobs they are asked to perform. Training should provide practical experience and practicality that complements the technical knowledge that is generally gained from attending learning institutions. Recognizing that early career technical professionals are not generally trained on how and when soft skills are appropriate for use to ensure success of their business encounters, leaders should ensure that there are the appropriate soft skills available and professionals are trained. Generally, organizations are attentive to ensuring that technical professionals receive the skills needed to render them technically competent. However, the same can't be said about the soft skills needed for success. Soft skills are recognized more today as being critical to the success of professionals in a technology-based environment. Some of the training that should be considered for all professionals include:

- Effective communication
- Implicit bias
- Diversity and Inclusion
- Handling difficult people
- Leadership
- Effective listening

The list above represents some of the skills that have been noted to be lacking among many technical workers. In order to cultivate and nurture workers so that they can reach their highest potential, other training areas that are considered key are critical thinking and strategic thinking skills.

When developing environmental training programs for an organization in all cases, it should be tailored to the group that is being trained. Providing too much information during training can defeat the purpose of educating individuals in the aspects that are required of them to perform their jobs safely and compliantly. Overexposure to information that is not pertinent to performing work safely and compliantly can increase information overload, confusion and loss of knowledge retention. Environmental regulatory requirements are generally massive for a company and can be vied in totality as overwhelming of individuals to grasp all of the requirements.

7.2.2 Training: A Necessary Investment for Success

An important element that can determine whether an organization will be successful in accomplishing goals and enjoy long-term sustainability is the access they have to the skilled workers needed to support their businesses. In order to ensure that the workforce has the knowledge needed, management must invest in their employees to ensure knowledge enhancement and sustainability. Knowledge retention and training strategy should be included in the organization overall business strategy. The training strategy must include the investment needed for success. What are some of the key benefits of investments in training? If closely evaluated, some of these benefits would include the following:

- Represents a potential competitive advantage for a company
- Increases employee confidence in their abilities and capabilities in accomplishing their jobs on behalf of the company
- Increases employee retention
- Improves and retains a high level of employee performance
- Consistency in performing work and responding to issues
- Closing the gap on employee weaknesses
- Increases innovation and strategic thinking
- Creativity in problem-solving
- Increases productivity
- Performance of work safely and compliantly
- Improved quality in products and services

A comprehensive training program does not automatically occur. It requires strategic planning, budget and resource allocation and management support. It is not always easy to determine the return on investment (ROI) when it pertains to training. However, when referring to compliance with environmental rules and regulations, it is easier to evaluate and compute the ROI value. For example, consider the following scenarios.

Scenario 1: Company A has been contacted by the Environmental Protection Agency that an inspection of their facility will take place within two days. The inspection will focus on waste handling and disposal. The company developed and implemented a comprehensive training program to ensure that workers understand the regulatory requirements associated with handling the waste that is generated as a result of performing their job responsibilities. The timing of the inspection was communicated to the workforce and work continued as usual. Two days later, the inspectors entered the facility and conducted the inspections and issued their final report one week later. The report was complimentary about the workers and their knowledge and adherence to hazardous waste regulations. The final report noted one violation during the inspection involving the labeling of one container found in a remote area, which was corrected on the spot. Although a violation was noted, the agency elected not to impose a financial penalty for the violation.

Scenario 2: Company B has been contacted by the Environmental Protection Agency that an inspection of their facility will take place within two days. The inspection will focus on waste handling and disposal. Management attempted to reschedule the inspection because they believe that they were not prepared for the inspection as they had not taken the time to fully train workers on the applicable portions of the regulations. A decision was made to reduce training cost by deferring training on the environmental aspects of job responsibilities with the intention of evaluating the associated cost for potential inclusion in next year's budget. The inspection took place two days later as planned and the final report issued to management. The report noted many violations, lack of knowledge, lack of leadership involvement and accountability and inadequate training of workers. The final report noted 40 violations that were accompanied by a financial penalty of $75,000. The regulators communicated to management that the fines could have been much more than they were assessed.

The two scenarios clearly demonstrate the ROI for companies that provide appropriate and timely training to workers. In situations that are noted for example in scenario 2, the cost of training could be much less than the financial penalty assessed. Other costs that are associated with receiving notices of violations can include the company obtaining a reputation of not protecting workers and the environment, frequent visits from the regulatory agencies and lack of trust from their stakeholders and the communities in which they conduct business. Training programs are necessary and should be included in the overall business strategy for a company to ensure the goals of the company are achieved and quality of product and services is delivered at the highest level.

7.3   APPLIED LEARNING

1. Define environmental justice and why the concept is necessary.
2. Summarize the role of environmental justice and the role it plays in helping to maintain a safe place for people to live and work.
3. Review the two scenarios in Section 7.2.2 and discuss the following:
   a.   Why is training important?
   b.   How can training be used to ensure compliance with regulatory requirements?
   c.   Discuss the differences between the leadership team in both scenarios. Include positive as well as negative attributes.
   d.   Discuss the culture of the organizations represented in both scenarios.
4. What steps can be taken to enhance knowledge retention?
5. List at least ten key benefits of implementing a training strategy.
6. List and discuss at least two applied learning methods.
7. List and discuss at least four active learning activities that can be used to facilitate critical thinking.

REFERENCES

Cole, Luke W., & Foster, Shelia R. (2001). *From the ground up environmental racism and the rise of environmental justice movement.* New York: New York University Press.
Kim, Kyoungna, Sharma, Priya, Land, Susan M., & Furlong, Kevin P. (2013). Effect of active learning on enhancing students critical thinking in an undergraduate general science course. *Innovative Higher Education, 38,* 223–35.
Markowitz, Gerald, & Rosner, David. (2002). Corporate responsibility for toxins. *The Annals of the American Academy of Political and Social Science, 584,* 159–74. Accessed March 17, 2020. www.jstor.org/stable/1049774.
Rhodes, Edward Lao. (2003). *Environmental justice in America, a new paradigm.* Bloomington, IN: Indiana University Press.

8 Requirements Tracking

Requirements in environmental compliance are not always limited to those stemming from environmental regulations put in place by regulatory agencies. There are often requirements associated with company-specific directives, industry standards, adopted best management practices, operational requirements, contractual requirements and memorandums of agreement, to name but a few examples. A cornerstone of any successful and sustainable environmental compliance program is the ability to identify and track requirements. From a regulatory compliance standpoint, a company's ability to maintain compliance is directly related to their ability to track and stay current with all applicable requirements.

8.1 KEEPING UP WITH NEW REGULATIONS

Since the environmentalism movement of the 1970s, environmental regulations have been constantly evolving. This evolution has been focused on the protection of human health and the environment while trying to keep pace with advances in industry and technology. Although there has been a concerted effort to protect the environment for over three decades, the regulatory setting in the United States continues to be very dynamic as new regulations are proposed and enacted constantly. Companies can proactively minimize compliance risk by establishing a method for staying abreast of new and updated regulations. There are basically two aspects that need to be considered for tracking new regulations: Who is responsible and how will tracking be accomplished.

The responsibility for tracking new regulations is of course a decision that needs to be made based on available resources and the number of regulations that apply to a company or organization. However, companies both large and small can benefit from establishing a requirements tracking and management system which includes identified roles and responsibilities for tracking regulations. The concept of a requirements-based management system is centered on identified requirements (regulatory, corporate, industry standards, etc.), the company procedures and other documents used to implement regulations and requirements and the personnel designated as responsible for identifying and keeping track of new requirements under their area of responsibility. The basic elements of a requirements-based management system are as follows.

As depicted in Figure 8.1, there is an organizational unit that is assigned the responsibility for a specific set of regulations or requirements. These organizational elements are the 'regulatory responsibility areas' referred to in Chapter 5 on requirements flow down. The regulatory responsibility area assigns personnel with the responsibility of tracking specific regulations. Depending on the number of regulations or other requirements that must be tracked, this can be a single person within the regulatory responsibility area or it may be multiple personnel, each tracking a specific regulation or set of regulations. These individuals are typically personnel who have knowledge or expertise (SMEs) regarding specific regulations. The SMEs

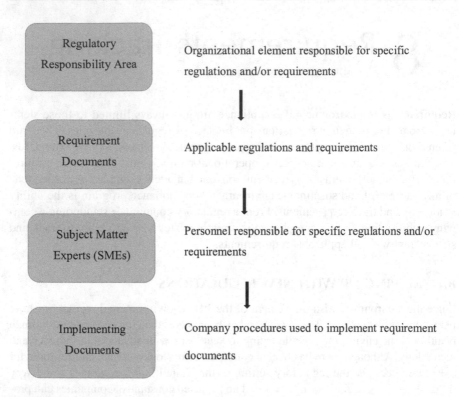

FIGURE 8.1   Basic elements of requirements-based management system.

are responsible for monitoring and tracking new regulations and changes to existing regulations within their assigned area of the regulations. As previously discussed, the SMEs would be responsible for communicating any new or updated regulations through established company channels so that implementation strategies and documents/procedures can be developed as appropriate.

To illustrate a specific example of how this concept would work, we will use the environmental organizational element that is the regulatory responsibility area for environmental regulations within Company ABC. In this example, we track a new regulation under the RCRA regarding the management of hazardous waste pharmaceuticals. In Figure 8.2, each gray shaded box is a basic element of a requirements-based management system and as such would be information that is documented or captured in an electronic database as the basic information related to the requirements associated with the management of hazardous waste pharmaceuticals. The elements are the responsible organizational element, the requirements document, the responsible SME and the implementing document.

The earlier a company is aware of a pending regulation that could affect their operations, the more time they have to develop an implementation strategy that is most efficient and best fits their operational requirements. It goes without saying that having knowledge in advance allows for a more proactive approach toward implementation versus learning of a new regulation after it is already in effect and having to reactively respond in order to stay in compliance. There are a fair number

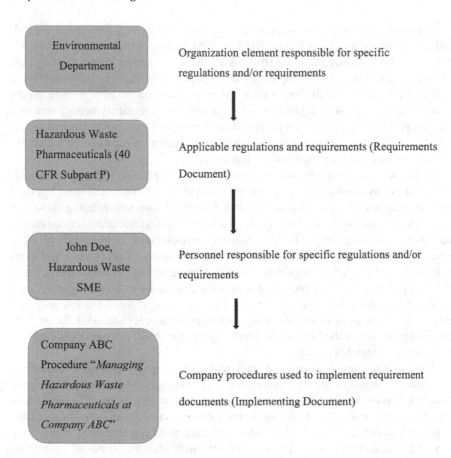

FIGURE 8.2 Requirements-based management system example.

of resources available for staying informed on proposed regulations and assisting companies' efforts to track new regulations and be better prepared to respond.

The primary resources for tracking new regulations are the Federal Register, regulatory agency websites and congressional and state legislation websites. Almost all of these information resources are available online and/or through subscription services. From an environmental regulation standpoint, the Federal Register is a great resource. The Federal Register is the official daily publication for rules, proposed rules and notices of federal agencies and organizations, as well as executive orders and other presidential documents (U.S. Government Publishing Office 2019). Every proposed federal environmental regulation must be published in the Federal Register as a proposed rule offering the public opportunity to provide written comments to the regulatory agency proposing the regulation. Because of this, the Federal Register is one of the best resources available for tracking proposed regulations from the time they are first proposed, through the comment period and to the final rule that will become regulation. The Federal Register also includes preamble discussion that provides insight into the reasoning behind the regulation and the intent of the regulatory agency behind each section of the proposed regulation. This type of information can be invaluable to companies when determining applicability and developing implementation strategies.

Other valuable resources can be found on regulatory agency websites at the federal, state, county and local levels. The U.S. EPA website will normally have information available on any proposed rulemaking that is occurring within their agency. As an example, if the EPA is proposing a new rule, not only will it be published in the Federal Register, there will also be information available on the EPA website. At the state level, there are state legislative websites and state regulatory agency websites that can provide information on proposed environmental legislation. Both also provide the ability to track proposed legislation from inception through the legislative process and to the final passage of the legislation.

There are also regulatory update services that are provided through subscriptions that are available. These can be very useful as the effort and time required for searching active legislation is performed by the companies offering these services. This can be an efficient way to save valuable time and resources for companies that operate in complex and highly regulated environments. Subscription services typically provide fairly detailed informational summaries of proposed regulations or legislation that are sent regularly (monthly, quarterly, etc.) to subscribers. These regulatory update services provide a means for companies to receive updates on a regular basis which can easily be reviewed to determine if there are any proposed regulations that may be of interest. Proposed regulations or legislation that may be of interest can be researched further as necessary by contacting the appropriate regulatory agency or legislative website.

Using any of these resources to track and monitor newly proposed regulations can assist a company's ability to stay ahead of the compliance curve by facilitating proactive approaches for developing implementation strategies prior to regulations actually going into effect. Utilizing some combination of these types of available resources can optimize a company's ability to track and keep abreast of new regulations in a manner that is most appropriate for the individualized needs of a company.

8.2 TRACKING AND IMPLEMENTING CHANGES TO EXISTING REGULATIONS

Tracking changes to existing regulations is of equal importance to the task of keeping up with new regulations. The advantages of changes to existing regulations are that companies have typically already identified the regulation and set about implementing the procedures or actions necessary to comply with the regulatory requirements. Evaluating the applicability and impact of changes to an existing regulation is usually less arduous than evaluating all of the requirements in a new regulation because obviously the effort can be mostly focused on the changes. Implementation is also somewhat easier as it commonly involves a revision to existing company procedures or other implementing documents rather than the development of an entirely new company procedure. However, as with new regulations, changes to existing regulation can represent a new requirement or set of requirements that require some type of action for compliance. A requirements-based management system as discussed in the previous section can easily be structured to accommodate the tracking and implementation of changes to existing regulations.

Expanding on the example presented in Figure 8.2, where 40 CFR Subpart P, *Hazardous Waste Pharmaceuticals*, was identified as a new regulation, let's assume that after the regulation became effective and Company ABC implemented their procedure: *Managing Hazardous Waste Pharmaceuticals at Company ABC*, EPA makes a change to the existing regulation. 40 CFR Subpart P includes 40 CFR 266.500 through 266.510 and for the purpose of this example, we look at a change to 40 CFR 266.507, *Residues of Hazardous Waste Pharmaceuticals in Empty Containers*. Figure 8.3 illustrates the information that would be captured in the requirements-based management system for this change to the regulation.

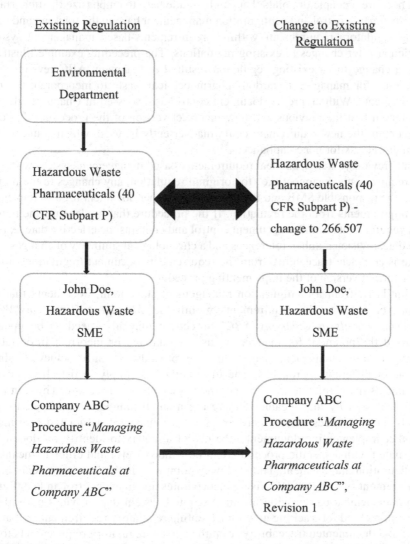

FIGURE 8.3 Capturing a change to an existing regulation in requirements-based management system.

The left side of Figure 8.3 depicts the elements that were captured in the require-ments-based management system for the original set of regulations in 40 CFR Subpart P. The right side of the figure captures the elements that would be associated with the change to the existing regulation, specifically 40 CFR 266.507. On the right side of the figure, John Doe remains as the SME since he is responsible for tracking 40 CFR Subpart P. The large double-sided arrow indicates that while there has been a change to 266.507 section of Subpart P, the remaining portions of the original reg-ulations still apply. The change to Section 266.507 of Subpart P results in a new revi-sion of Company ABC Procedure '*Managing Hazardous Waste Pharmaceuticals at Company ABC*,' as shown on the bottom right of the figure.

There are a couple of points that can be made here to emphasize the importance that document control and configuration management have in the tracking and man-aging regulatory requirements within a requirements-based management system, particularly for changes to existing regulations. The preceding example illustrates how a change to an existing regulation resulted in Company ABC revising their procedure for managing hazardous pharmaceutical waste to incorporate the new requirements. Without proper document control as discussed in Chapter 6, there is a potential that the previous and now incorrect version of the procedure that does not contain the new requirement could inadvertently be used, which could put the company at risk for noncompliance.

As previously stated, an ideal requirements-based management system will track the requirements document (i.e., the original regulation, any changes to the regula-tion), the responsible SME and the procedure or other document used to implement the requirements from the regulation. If the procedure that implements the regula-tory requirements is under document control and contains the effective date, expira-tion date, revision number, references and a chronological summary of changes, then there is complete traceability from the requirements document (regulations) to the most current version of the implementing procedure.

Similarly, from a configuration management standpoint, documents that ref-erence or point to the requirements within the '*Managing Hazardous Waste Pharmaceuticals at Company ABC*' procedure may also need to be updated to avoid the potential for inadvertent use of outdated or incorrect information. Configuration management of implementing procedures can be achieved using a couple of different approaches. The first method is simply to list all associated documents in the references section of the implementing procedure. This can work well and is easily manageable if there are a small number of associated docu-ments. For instances where there are a large number of associated procedures or other implementing documents, the use of a matrix to identify all documents associated with a specific procedure can be a more useful and efficient means for configuration management. The takeaway point here is that a requirements-based management system that effectively incorporates document control and configura-tion management can provide for an integrated system that provides not only an efficient method for tracking and implementing regulatory requirements but also a clear and documented traceability from the current regulatory requirements to the current implementing documents.

8.3 PROFESSIONAL WORKING GROUPS
IMPACTS AND PARTICIPATION

Participation in professional working groups, associations and societies is also a great way to stay current on regulatory issues. There are countless professional organizations that are focused on environmental topics, including hazardous waste, storm water, wastewater, air pollution and environmental sustainability, to name just a few. Many of these groups are not only focused on specific areas of regulatory interest but also promote training, education and certifications. These groups of environmental professionals can contribute greatly to the regulatory process by sharing specialized expertise, direct relevant experience and industry lessons learned with regulatory agency personnel and lawmakers at the federal, state and local levels. The process the EPA must follow when proposing and promulgating new regulations and changes to existing regulations requires that opportunities for public participation be provided (US EPA 2018). At the federal level, public participation mechanisms include publication in the Federal Register for public comment, EPA public affairs publications and brochures, agency websites and public meetings or workshops. Because they do represent a collection of highly knowledgeable personnel and significant cumulative experience, professional working groups can impart a significant amount of valuable input to the regulatory agencies. Regulatory agencies often consider these professional groups as essential resources for the regulatory rulemaking process. When professional working groups are successful at establishing a constructive working relationship with regulatory agencies, their input is often solicited giving them the opportunities to positively influence the regulatory language and content. Professional environmental groups and organizations are quite often very actively involved with state, county and municipal regulatory agencies and the respective rulemaking decision processes. Participation in these types of professional environmental organizations can be of great benefit to individuals as they offer great opportunities for professional growth and networking. These professional groups can also be beneficial to the companies that have employees who are participating members by providing access to a collection of knowledgeable experts and resources that can provide up-to-date information, including regulations, technology, lessons learned, industry trends and professional contacts.

8.4 CONVEYING LEGISLATIVE CHANGES

Tracking proposed legislation and legislative changes that are related to environmental regulations can afford companies the opportunity to begin formulating changes in implementation strategies that may be necessitated by an upcoming new or revised regulation. As previously discussed, this could be legislative changes that are being proposed at the federal, state or local levels. Conveying proposed legislative changes as early as possible to relevant managers and stakeholders will allow the opportunity for companies to assess the potential impacts relative to resource needs, operational changes, operating costs, etc. Communicating proposed legislation as soon as the information is available and then providing periodic updates to managers and stakeholders will enable the company to adjust initial implementation strategies as the proposed legislation or regulation undergoes revision as it progresses through the legislative process. This can be a very beneficial planning tool that can forecast potential increases in cost and resource commitments as well as compliance obligations.

A graded approach can be applied when assessing proposed legislative changes in that those changes which pose potentially significant impacts to a company can be monitored more closely with an increased focus on preparatory actions. Initial preparatory actions could include participating in the review and comment of proposed legislation or regulation if opportunities for public comments exist. A good approach that can be applied here is to obtain input from relevant stakeholders within the company and provide a consolidated set of comments to the agency or legislative body sponsoring the legislation. As discussed above, this is also a great opportunity to take advantage of professional working groups that may have direct involvement with the regulatory agency or legislative body. In addition to public participation actions, other preparatory actions that can be taken related to implementation planning include:

1. Developing initial implementation strategies, phases and timelines
2. Establishing roles and responsibilities
3. Evaluating personnel resource needs
4. Evaluating cost impacts and budgetary needs
5. Assessing operational changes
6. Identifying company procedures and/or documents that may be needed

This is not intended to be an exhaustive list, but provides some examples of the types of actions that can be taken in advance when legislative changes and proposed regulations are identified early in the process and properly conveyed to company managers and stakeholders.

8.5   APPLIED LEARNING

Review the case study and respond to the questions that follow.

8.5.1   CASE STUDY

As the new operations manager for Company ABC, you have been made responsible for developing a system for identifying and tracking all of the requirements that the company must comply with. In your role as operations manager, you oversee and direct the operations of the company. The organizational departments that report directly to your position include environmental compliance, health and safety, transportation, manufacturing and finance. Describe your approach for each of the following:

1. How would you assign roles for tracking requirements?
2. What responsibilities would be associated with these roles?

3. How would you propose to keep track of applicable requirements?
4. How would you track procedures or other documents used to implement requirements?
5. List the information that you think would be important to capture in your system for managing requirements.

REFERENCES

U.S. Government Publishing Office. (2019). Federal register. *Govinfo*, February 11. Accessed March 23, 2020 from HYPERLINK "http://www.govinfo.gov/help/fr" www.govinfo.gov/help/fr.
US EPA. (2018). Laws and regulations. *US EPA*, November 26. Accessed March 26, 2020 from https://www.epa.gov/laws-regulations/get-involved-epa-regulations.

9 Assessing and Measuring Program Performance

9.1 DEFINING SUCCESSFUL PERFORMANCE FOR A COMPLIANCE PROGRAM

A company performance has a direct impact on profitability, compliance and sustainability. As a part of achieving success, a company must also ensure that they comply with all of the regulatory requirements that are applicable to their business. To do so, leaders must have a working knowledge of the status of the company through predetermined mechanisms such as performance measurement. Measuring performance involves measuring the actual performance output against its intended goals and the important attributes that are necessary to achieve the goals. Performance measurement requires collecting, analyzing and reporting of data that represent the performance of an organization, system, processes, practices or individual. This requires a top-down approach by management for setting performance criteria, measuring outcomes and instituting changes as needed. To ensure that performance measurement is a core business activity, the measurement process and protocol should be included in the company's overall strategic plan. The strategic plan should provide performance targets for the organization which sets the direction for operation and performance for each group and member. Performance measurement is a vital part of the process of monitoring that should demonstrate how an organization is progressing in various aspects of operations. The following types of questions should be answered or at least explored by performance measures developed in order for them to be deemed as being useful:

- How is the organization performing with regard to its goals?
- Does the organization have the resources needed to accomplish goals (people, financial, equipment, etc.)?
- Are the products developed aligned with the approved specifications?
- Does the product deliver high quality?
- Are the needs and expectations of customers met?
- Are regulatory commitments being met?
- Is worker training maintained to meet the current and future needs of the organization?
- Is there a process in place to support continuous improvement?

The aforementioned inventory only represents some of the questions that performance measures can and should answer for organizations. If designed and used appropriately, performance measuring can save a company in revenue expenditures, loss of business opportunities and negative impacts on their reputation. Many organizations continue to struggle with identifying, developing and implementing meaningful performance measures that provide an ongoing view of exactly how the company is performing. As such, they develop metrics to drive improvement, provide a focus on what is deemed to be important by the senior leadership team and measure the overall performance of an organization. Metrics that are developed are in the form of leading or lagging indicators. Both types of metrics provide some useful information, however; only leading indication can influence future performance, whereas lagging indicators can be used to analyze past performance. Understanding the past through lagging indications can help management plan for the future. Leading indicators are not always easy to identify and measure; as such, there tends to be a tendency to utilize lagging indicators more freely. A well-balanced performance measurement program should contain a healthy balance of both with a greater focus on leading indicators. This struggle to develop metrics that accurately depict performance is real and continues today. Some issues associated with the development of metrics include the following:

- There are too many of them.
- Metrics that are backward-looking with no account for what will happen in the future tend to be developed and cited more.
- Little or no forward-looking elements.
- The same level of importance given to all metrics.
- A false sense of security and achievement when they are not measuring what's important.
- Metrics are not representative of the business in which it is deployed.
- They are not reviewed and used in business decisions.

Developing a comprehensive monitoring program to gauge compliance with environmental regulations can be challenging for many reasons. Depending on the business profile, a company can be subjected to many regulations that must be monitored. This process takes time and, of course, money: Just how much time and money it takes depends on a variety of aspects that go beyond just the regulations that must be adhered to. It also includes the staffing requirements to aid in compliance, record-keeping and knowledge required to be maintained. The process used to develop an environmental monitoring program should include someone knowledgeable in environmental regulatory requirements to inform development of the monitoring process.

When developing a comprehensive monitoring program, there are three types of information that should be explored and included in the monitoring process decisions. The three types of information are represented in Figure 9.1. Together these information components provide a thorough view of performance that can impact past, present and future performances.

Qualitative measures: The raw data represented by qualitative data are in words where observations and comments are evaluated to draw conclusions. Surveys,

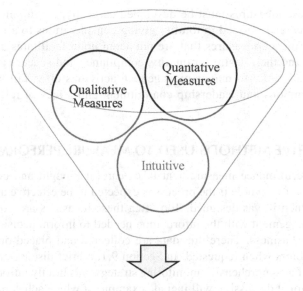

Comprehensive Performance
Measures Components

FIGURE 9.1 Performance measures.

interviewing workers and clients, benchmarking and focus group discussions are
often used to gather qualitative data to uncover viewpoints to inform performance in
a specific area. Metrics developed based upon qualitative data exhibit a level of subjectivity that may not be recognizable and can impact the accuracy of the results and
interpretation of performance. In short, qualitative data provide feedback or characterization of information.

Quantitative measures: The raw data represented by quantitative means are
numerical in form. This form of data is favored because it is generally easy to gather,
analyze and explain as it involves counting or applying a numeric value to an object
or event. In short, quantitative data can be counted.

Intuition: It can be used to a lesser degree (not recommended as a strategy); sometimes intuition can provide a feeling of uneasiness or speculative that something is
not right. These feelings that something is not right should be evaluated to ensure
that there are not any underlying issues that can prevent performance or the culmi-
nation of accurate performance indicators. Some refer to intuition as nonconscious
or instinctual; regardless, this mode of thinking can be of use on a limited ad hoc
basis. This should, however, not be relied upon as the only method to determine
performance.

It is important to keep in mind that performance measures can drive behaviors
that will assist in achieving results as well as drive behaviors that will derail
accurate measurements and reporting because of a focus on the end product.

Therefore, these measures must be developed carefully, communicated, openly and thoroughly, and utilized in totality giving considerations to all information. Having performance measures that are not acted upon facilitates a loss of trust in leadership and their desire for quality, compliance and success and is useless to an organization. Performance management activities often fail when there is a lack of champions and leadership commitment and a lack of relevancy to the business.

9.2 EFFECTIVE METHODS USED TO MEASURE PERFORMANCE

Developing performance measurement takes careful thought and considerations for the business as a whole if the process is expected to be effective and serves the purpose in which it was designed. Too often these measures are not effective in providing management with the information needed to inform decision on various aspects of the business. Therefore, data are collected and placed on the shelf to present to auditors when requested. In Section 9.1, a brief discussion of the three components of a comprehensive monitoring strategy was briefly introduced. In this section, additional discussion will include example of what each component comprises and some examples or suggestions on interpretation of the results generated (Tables 9.1 and 9.2).

The most effective measures to use when measuring the performance of the environmental compliance program are different for different organizations even if they are in the same market. Therefore, performance measuring is organization specific. However, this does not mean that benchmarking other organizations to seek best practices is not a good strategy before developing a strategy, monitoring protocol and metrics. The caution is to ensure that the protocol meets the needs of the organization that is being monitored.

TABLE 9.1
Data Collection

Component	Collection Methods	Sources Examples
Quantitative	Interviews	Strategic well-structured questions resulting in numerical data
	Surveys/questionnaires	Questions design to yield responses that are numerically based
	Observation	Observations structured to yield numeric data type
	Document review	Observing and reviewing specific information that can be cataloged numerically
Qualitative	Documentation review and analysis	Reports, procedures, policies, newsletters, work control documentation, job hazard analysis
	Observations	Observe work being performed, interactions between workers and leadership, leadership involvement in work activity
	Interviews	Workers, management, customers, stakeholders
Intuitive	None specific	Instinctual with no formality

TABLE 9.2

Data Analysis and Presentation

Component	Methods	Examples
Quantitative	Graphs and charts	*Line graph* – effective in depicting more than five data points over a period of time
		Bar graphs (clustered) – for displaying several series of data
		Pie charts – works well for general information or findings and demonstrates parts of a whole
		Tables – tabular presentation for data comparison
		Stem and leaf plot – provides a mechanism to list data in a compact form
Qualitative	Graphs and charts	*Pie charts* – works well for general information or findings and demonstrates parts of a whole
		Text – a brief written description
		Pareto diagram – allows for comparison of items
		Bar graph – arranged in order of frequency to allow easy identification of important characteristics
Intuitive	Feelings and assumptions	Instinctive feelings may be discussed only with others in general terms

9.3 IDENTIFYING WHAT IS IMPORTANT: CORPORATE AND STAKEHOLDER PRIORITIES

Before identifying what is important to measure, it is just as important that the leadership team and organizational members understand why measurement is necessary. In fact, it is important to have members of the organization onboard with the concept of measuring performance and the process that will be used to collect, analyze and communicate the data. It is critical to have the necessary parties involved in collecting the data and providing them to the assessment or monitoring team without concern of retaliation if the data do meet the expectation of leadership. The importance of performance measuring resides in the knowledge that it improves productivity, reduces costs, aligns activities with the business plan, facilitates identification and implementation of best management practices and improves performance of an entity. The measurement of performance is a continuous improvement process which involves checking the performance against the standards that demonstrate compliance that was developed to be followed by the entire group or organization. Monitoring performance begins with some thought-provoking questions. Some of these questions are:

1. Why measure?
2. What is important to accomplishing the goals of the business?
3. What is important to customers and stakeholders?
4. What does success look like for an organization?
5. How should and can measuring be conducted?

The responses to these thought-provoking questions and others form the basis or foundation on which to build the performance-monitoring program and the type of metrics that should be developed, deployed and monitored. In identifying what to monitor and how to monitor, special attention must also be paid to the needs of the customers and stakeholders. In addition, the answers to these questions will more than likely spark more questions that will be useful in supporting the importance of monitoring that can be used to introduce and sell the concept along with the process to the organization as a whole. It can also demonstrate to customers and stakeholders that the company is responsible and interested in ensuring their needs are met and quality is maintained. The method used to determine what to monitor should also be utilized when addressing environmental regulatory requirements with the inclusion of the requirements specified by regulations and any permits or consent agreements held by the organization.

9.4 EVALUATING ENVIRONMENTAL COMPLIANCE PROGRAMS AND PROCESSES

Environmental compliance programs are developed and documented in organizations to ensure that members are aware of their responsibilities regarding the expectations for following environmental rules and regulations and how compliance can be achieved. A documented program also indicates to the regulators and auditors that you intend to be compliant and have a process for your workers to share in the compliance burden for the company. These programs are documented in the form of policies or/and procedures. The policies and procedures outline the environmental regulations that are applicable to the business and the process and practices used to ensure compliance and safety of workers and the environment. One should be cautioned that developing a program that is not followed can be perceived as being out of compliance by auditors and regulators and an organization can be penalized for what is documented and not followed or implemented. A penalty can be imposed even if it is not a regulatory requirement.

A comprehensive assessment program is a key component of a demonstrated effective compliance program. Assessing your program is necessary because regulatory requirements change frequently and as such the organizations' internal document program will change as well. Having a current program that is documented and understood by organizational members is important to the program being embraced and followed. In assessing the environmental compliance program (the documentation), some elements are important to include in the review.

The first basic step to assessing a company's program is to perform a crosswalk between the regulation and the internal organization procedures for all laws and regulations that the organization is expected to maintain compliance. This review is to ensure that the documentation is in alignment with the regulations. The assessment can feel like a daunting task, depending on the amount of applicable regulations that must be followed. As such, it may be a good idea to develop an assessment schedule for assessing the various regulations and not attempt to complete the task for all regulations at once. Key areas to evaluate include:

- Identify any changes made to regulatory requirements that are not included in internal procedures.
- As a result of regulatory changes, is there a need to change or develop new training for employees?
- Identify and ensure the workers needing training have the training requirement listed to their job demands profile.
- Are procedures and policies clear, easy to understand and followed?
- Are there any connected or referenced internal documents needing change as a result of changes in regulatory requirements?

9.5 EVALUATING IMPLEMENTATION OF ENVIRONMENTAL REQUIREMENTS

Implementation is where one would say 'the rubber meets the road.' Improper or inadequate implementation can result in a significant amount of noncompliances and financial penalties imposed on organizations. There are several methods that can be used to evaluate the effectiveness of an environmental program. These methods involve taking advantage of internal resources and utilizing external resources having the appropriate level of expertise to provide program insight.

Internal: Each organization should have an effective assessment program, one that is designed to evaluate a program or process effectively and uncover inefficiencies and noncompliance. A comprehensive assessment program should include a process that allows for identification of issues, resolving issues, validating that solutions are appropriate to prevent reoccurrences, tracking and trending of issues, resolving of issues and a comprehensive lesson-learned program.

External: Inviting external agencies into an organization to assess performance is a good way to learn from the viewpoint and experiences of others. External agencies are often viewed as impartial and have the capability to report honestly on the status of an organization, outline the issues that need to be addressed to improve efficiency and communicate accurately without apprehension compliance posture and areas needing improvements.

Implementing environmental requirements in organizations is not an easy task, especially when the requirements are many. An organization must employ several means to inform them of their implementation burdens and status and to identify areas for improvement. Some of the ways that environmental requirements can be evaluated are:

- Through assessments
  - Internal workers with knowledge in the regulatory requirements
  - External agencies hired by the organization
  - Regulatory inspections (least preferred method)
- Feedback from workers
  - Focus groups
  - Interviews
  - Surveys

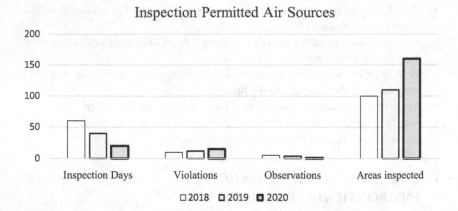

FIGURE 9.2   Regulatory metric example.

The objective is to use all forms of assessments or evaluations in totality when evaluating a program and avoid focusing singularly on one piece of data. Focusing on one data source can lead to a false sense of security of 'things are on fire.' Data sources must be analyzed objectively to uncover the true status and development of the path forward. The graph in Figure 9.2 shows results for inspection of permitted air sources for three years performed by an internal team.

A point of note is that although an inspection program is in place and implemented, violations are still noted during the internal inspections at an increased rate yearly. When analyzing the graph in Figure 9.2, the following can be gleaned:

- The organization has an inspection program for permitted air sources that has been implemented for the past three years.
- A significant amount of the area is being inspected each year.
- Areas inspected increased over the years.
- Areas inspected increased as inspection days decreased.
- Conditions that can lead to a violation are not significantly observed.
- Violations are increasing over the years.

Just as important as having a process to evaluate performance, it is important to utilize the data and continue to improve programs and processes. Oftentimes organizations collect data and place them on the shelf or fail to use them appropriately. In order to improve the program and the results obtained from assessing permitted air sources, the following should be considered and explored:

- Does the inspection program for permitted sources need to be revamped? Are the correct metrics being used? This should be the last question asked, although they are presented as a first point to explore. One caution here is that don't assume that the program or process is the problem and the data or the information discovered is not telling the correct story.

- Why is the amount of violations increasing each year albeit slightly? The first thought or question should be: Is program implementation decreasing? Effective program implementation should yield continuous improvement and there should be a decrease in violations. However, degradation in implementation can yield the results seen here.
- Are the right areas being inspected? The key is to ensure the correct areas are being inspected at the appropriate frequency – areas and sources where work is subjected to the requirements of the permit.
- Are the right sources being inspected? Ensure the appropriate components of the source are inspected. These components will include the parts of the source that contribute to or in some way impact emissions of a chemical or hazardous substance. Ensure that those critical parts are maintained and documented at appropriate frequency.
- Are the people performing the inspection knowledgeable? Ensure that the workers responsible for assessing programs and processes are knowledgeable of the operation parameters and the regulatory requirements. There are times when performing the assessment needs more than one individual to ensure that the appropriate level of knowledge is devoted to completing the assessment comprehensively.
- Is enough time allowed to conduct a meaningful assessment? Note that the inspection days decreased while the number of areas inspected increased may indicate that there is a rush to complete a task that is on the books. In such cases, the quality of the process is being impacted and perhaps quantity of inspections performed has taken precedence over program improvement.

9.6 APPLIED LEARNING

1. What does a successful compliance program look like for a large organization? A small organization?
2. List the comprehensive measures components. Explain each in detail.
3. Which component of the performance measures components is most desired and why? Which component is least desired and why?
4. Explain how to identify the important aspects of a business that should be monitored.
5. What methods should be used to evaluate environmental program effectiveness and compliance?
6. What methods should be used to evaluate implementation of environmental program effectiveness and compliance?

10 Tracking and Trending Performance

Tracking and trending how well an environmental compliance program is performing is an essential and ongoing necessity for a sustainable compliance program. An overarching characteristic of a high-functioning compliance program is the ability to continually adjust and make improvements based on performance. Tracking performance issues and trends can provide invaluable insight into program aspects that may be experiencing compliance issues or program elements that are not in alignment with the company mission or established strategy. In order to track performance that is meaningful to the compliance program and the company, relevant metrics must be identified and measured in a manner that enables a company to make focused adjustments to program aspects or processes that are not meeting company expectations, regulatory compliance obligations or external stakeholder expectations. The actual metrics and performance indicators to be used to measure a company's environmental performance will most certainly be based on the specifics of each individual company and associated circumstances. Identifying the right set of metrics can be challenging and most often is an iterative process that requires refinement over time.

10.1 IDENTIFYING AND USE OF MEANINGFUL METRICS

Meaningful metrics for environmental performance can cover broad topical areas and are not necessarily strictly limited to regulatory compliance and the number of violations a company receives. Although regulatory compliance and environmental violations are certainly important elements of environmental compliance as performance indicators, there are many other aspects that should be considered when determining which performance metrics are best suited for a company or an organization. Many operational and functional aspects that are not directly responsible for environmental compliance can influence environmental performance. Business functions such as supply chain management, research and development, manufacturing and product testing, to name but a few, can all include elements that can interact with the environment in some way. Identifying those functions within a company that interacts with the environment, associated indicators for performance and a set of metrics that can be used to analyze performance data will provide the tools companies need to continually improve their environmental programs. For the purposes of this book, we consider metrics to be a measure of an activity at a specific time and are used in conjunction with performance indicators which encompass objectives toward specific targeted goals (Walczak 2014). From an environmental compliance program standpoint, there are three broad areas where metrics can be evaluated for performance: how well requirements are flowed down into the organization, environmental reporting and alignment with the corporate strategy. Within each of these broad categories, there are elements that can be utilized to monitor performance by identifying appropriate indicators and establishing metrics that capture the information that a company has determined to be relevant measures of performance.

10.1.1 How Well Are Regulatory Requirements Identified and Flowed Down

Chapter 5 discussed the importance of identifying applicable environmental regulations and requirements and flowing them down into the organization. The flow down of requirements can be accomplished with consistency if a defined process for communicating these requirements is established and utilized. There are several aspects of requirements flow down that lend themselves to ongoing periodic evaluation to assess how well requirements are being identified and communicated within the organization. Some suggested elements of the requirements flow down process that can be utilized for performance metrics are:

- Timely identification of regulatory requirements
- Stakeholder review process
- Implementation strategy development
- Procedure development and control process
- SME training and qualifications
- Requirements tracking process

This is not intended to be an exhaustive list, but provides examples of some of the elements associated with requirements flow down that could directly impact environmental compliance if they failed to function as intended. Metrics could be established for each of these elements, depending on the needs of the company. As an example, if an audit or a regulatory inspection identified that a company was operating under an outdated regulation, then an evaluation of each of these elements to determine the causal factor(s) is warranted. Once the problem area is identified, it can be monitored against an established set of metrics so that actions can be taken to prevent recurrence. If in our example the cause was regulations not being identified in a timely manner, then it might make sense to establish a set of metrics that capture such things as the number of applicable regulations and time frames from proposed regulatory information first becoming publicly available to when it is identified within the company as applicable and communicated to internal stakeholders. Measuring how well the requirements are being flowed down within the organization is often a process that is overlooked until a company is forced to react due to a regulatory inspection, notice of violation or related noncompliance issue. Due to the dynamic nature of environmental regulations and the importance of staying up to date with current regulations, a viable compliance strategy will benefit from periodically evaluating and measuring how well their process for identifying and flowing requirements is performing.

10.1.2 MEETING REGULATORY REPORTING REQUIREMENTS

A company's environmental reporting requirements can include a wide variety topics and regulatory obligations ranging from water discharge monitoring reports to hazardous waste facility biennial reporting to Toxic Release Inventory (TRI). A straightforward measure of performance relative to environmental reporting that is directly related to compliance is simply whether the reports are meeting the regulatory requirements. In other words, are they being submitted on time and does their content meet the regulatory agency expectations? Environmental reports that are required by regulations that do not meet submittal dates or do not adequately address required content can place a company at risk for violations, fines and penalties. Companies operating in highly regulated environments can be responsible for a large number of regulatory reporting obligations that can include weekly, monthly, quarterly and annual reporting. In these cases, is it especially important that a company's process for meeting reporting obligations is periodically evaluated to assess how it is performing?

Environmental reports are in and of themselves reflective of how an operation, a facility or a company is performing to a specific set of requirements. As such, environmental reports offer a substantial opportunity for a company to measure performance at various levels. As an example, many companies, shareholders and external parties view a company's TRI report as one measure of how they are performing environmentally. Since the TRI report requires companies to report hazardous material inventories and releases above certain thresholds, an assessment can be made as to how well companies are managing their hazardous material inventories and their impact on the environment. Companies can utilize almost any environmental reporting to define and capture metrics that will provide for a measure of performance and enable process improvement. This approach can be used to track performance against expectations from external interested parties, corporate or internal expectations and can be an invaluable tool for continual improvement and managing risk. In their analysis of environmental risk and reporting, Hood and Nicholl (2002) report that 100% of respondents to an environmental risk assessment and management survey indicated that their companies had identified environmental risks with the main categories being water pollution, land contamination, waste issues, air emissions and regulatory compliance. Each of these categories almost invariably has some type of required report which further supports the assertion that environmental reports can provide great value in measuring performance, process improvement and managing environmental risk. Ideally, environmental performance that is measured and reported externally should be coupled to that being measured and reported internally so that both have connectivity to a company's process improvement mechanism (Scherpereel et al. 2001).

10.1.3 HOW WELL DOES THE PROGRAM PROMOTE COMPLIANCE AND THE CORPORATE MISSION

How well a company's environmental strategy aligns with the overall corporate mission is in many ways an extension of the company's level of motivation toward compliance and overall environmental management. A company's level of motivation toward environmental compliance and management can range from very low to very high, depending on a multitude of factors. The term 'motivation' is used here to define the degree to which a corporate strategy incorporates not only environmental compliance but other environmental considerations, such as recycling, sustainable purchasing, life cycle assessments, etc. Companies operating under minimal regulations may have minimal compliance risk and do not have a need to expend a great deal of effort on compliance. This, of course, does not imply that these companies cannot be motivated toward compliance and the environment, only that the compliance risk is smaller and perhaps more manageable. Companies operating in a highly regulated environment would most likely be faced with significantly more compliance risk and may focus a lot of effort and resources on compliance. At the same time, larger companies with significant compliance risk may also place a great deal of emphasis on their activities that interact with the environment and promote a holistic attitude toward environmental management. There isn't one approach that is best for every company. A company's motivation toward compliance and the environment as a whole is individualized and specific to each company's set of circumstances. Corporate motivational level toward the environment is really the type of culture a company wants to create and support within their organization. This is not simply how 'green' a company is but also a consideration of the attitude regarding compliance versus mission. In other words, does the corporate mission always take precedence over regulatory compliance, or does compliance take precedence over mission? Most companies that have to adhere to environmental regulations face that decision at some point and, ideally, they gravitate toward operating conditions where compliance is an integral part of the corporate mission and strategy. Many companies strive for regulatory compliance but are also driven toward a more holistic environmental strategy by the increased social demands for corporate environmental responsibility (Hood and Nicholl 2002).

In today's competitive corporate environment, many companies must consider these social demands and other external expectations from entities such as community leaders, shareholders and special interest groups. Companies can benefit from integrating these external expectations into their corporate strategy in parallel with their internal expectations. A company's performance relative to external expectations should help formulate the company's environmental strategy and correlate to internal performance expectations (Scherpereel et al. 2001). The desired alignment of a company's compliance program with the overall corporate strategy is initially defined by the expectations of the company managers and executives. As companies develop their strategies, decisions are made as to how environmental compliance will fit into the overall corporate strategy. As discussed above, this is based on the level of motivation the company has toward environmental compliance, the environment overall and the type of culture they wish to establish within their organization. How well the two are aligned can be determined by establishing metrics to demonstrate that the company's internal expectations for compliance are being met. Companies that must consider expectations from external entities can also establish metrics for how well they are meeting those expectations. Inherently, the higher the level of correlation between these two sets of expectations, the easier the development of strategies for tracking and measuring performance against both sets and making process improvements as necessary. Metrics can be used to communicate performance to both internal and external audiences; but in order for them to be meaningful, it is important to understand what the internal and external expectations are relative to performance and what the appropriate performance indicators are.

10.2 PERFORMANCE INDICATORS

Companies can begin identifying specific indicators for environmental performance by determining those activities or functions within the organizational elements of the company that can affect or interact with the environment. Environmental management systems such as ISO 14001 and ISO 14031 refer to these activities as environmental aspects (International Organization for Standardization 2015). Identifying the environmental aspects within a company can be pivotal in helping managers to better understand the types of environmental performance indicators (EPIs) that can be most beneficial for the needs of the company.

Determining environmental aspects is not a process bound by a set of hard and fast rules or prescriptive processes. Activities or processes being evaluated for environmental aspects can be effectively evaluated by considering the life cycle of the product or process. A life cycle analysis can provide information regarding environmental impacts and interactions throughout the life cycle of a product or process that can contribute to the corporate decision-making and process improvements (Van Der Vorst et al. 1999). As shown in Figure 10.1, a product or activity life cycle will commonly include material input, production, distribution, use and disposal. Life cycles can obviously contain additional elements and finer degrees of granularity, but

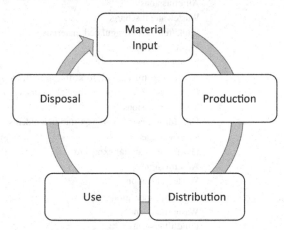

FIGURE 10.1 Common elements of a product life cycle.

this simplistic example is presented here in order to illustrate the concept of life cycle evaluation for determining environmental aspects. Using this example, each life cycle element would be evaluated for interactions with the environment. It is worth noting that environmental aspects do not have to be limited to negative interactions with the environment as there can be environmental aspects that are associated with positive interactions which can be beneficial to identify for performance tracking. Table 10.1 lists some examples of environmental aspects that could potentially be associated with each of the life cycle elements.

The potential environmental aspects listed are examples only and far from being an exhaustive list. Actual environmental aspects would obviously be dependent on the specific product and the activities associated with each element of the life cycle. Each of the potential environmental aspects listed presents a potential component for which indicators could be established for the purposes of measuring environmental performance.

The decision on which aspects to measure can depend on a variety of factors, including corporate strategy, management interests, regulatory risks and

TABLE 10.1
Life Cycle Elements and Potential Environmental Aspects

Life Cycle Element	Potential Environmental Aspect
Material Input	Hazardous materials
	Nonrenewable resources
	Renewable resources
	Recycled materials
	Sustainably produced
Production	Energy usage
	Water usage
	Air emissions
	Wastewater discharge
	Spills/releases of regulated materials
	Waste generation
	Land/space use
Distribution	Fuel usage (trucks, rail, forklifts, aircraft, etc.)
	Energy usage (conveyors, lifts, elevators, etc.)
	Vehicle emissions
	Hazardous materials transportation incidents
Use	Energy usage
	Hazardous material exposure
	Air emissions
	Waste generation
Disposal	Waste transportation
	Waste treatment
	Landfill capacity

expectations from external entities such as shareholders, community leaders and special interest groups. It would seem imperative that the indicators and metrics chosen to measure environmental performance are in alignment with the corporate strategy which ideally integrates and is congruent with regulatory risks, relevant external expectations and management's attitude toward the environment (Figure 10.2).

This type of congruency is presented by Scherpereel et al. (2001) in a study in which an analytical model is used to develop EPIs. The study identifies three factors that contribute to an environmental strategy: internal factors, external factors and the company's level of ambition relative to their environmental strategy. Other studies have shown that two primary factors come into play regarding corporate behavior toward the environment: first, the external factors (legal, social, political, economic); second, the mindset of management toward the expectations of each of these external factors (Kagan et al. 2003). These referenced works support a clear connection between external expectations (i.e., regulators, special interest groups and communities), internal factors (regulatory risk, organization structure, resources) and the corporate attitude and level of motivation demonstrating the importance of these factors in the identification of performance indicators. Companies can benefit from establishing clear and distinct environmental goals and objectives as a prerequisite to developing performance indicators (National Research Council 1999). This approach is advantageous as it is essentially important that the criteria for performance indicators have a strong correlation with the intended objectives of the indicators (Eyles and Furgal 2002).

Once performance indicators are identified, specific metrics can be established that capture the data and information needed for collection and analysis in support of internal and external communication of performance. Performance indicators, metrics, data collection and analysis should be part of a continual improvement process

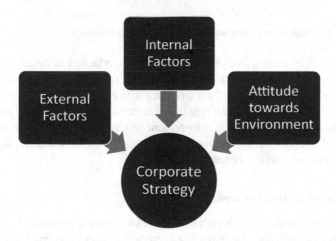

FIGURE 10.2 Factors influencing corporate environmental strategy.

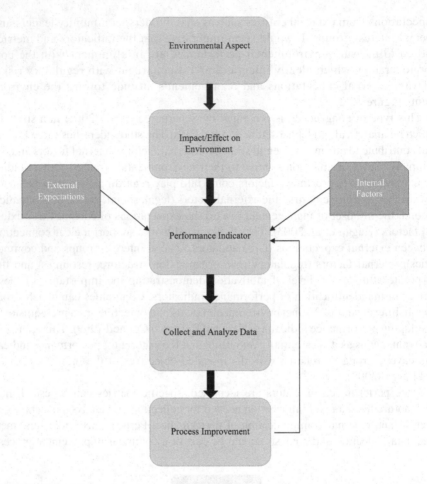

FIGURE 10.3 Process improvement and performance indicators.

loop where each of these associated elements can be adjusted as necessary to ensure performance measures are meeting external and internal expectations (Figure 10.3).

The collection and analysis of performance data can provide invaluable information that can help a company with regulatory compliance, meet internal and external expectations and manage risk. It is important when analyzing performance data to understand if it represents the evidence of something that has already taken place or if it is indicative of a future occurrence.

10.2.1 Leading and Lagging Indicators

Performance indicators can be developed and used for many different purposes, depending on the overall goals and objectives that a company desires to accomplish by a specific measurement of performance. There are indicators for input, output, response, process, operational, intermediate, financial, qualitative, quantitative and many others. Many of the various terms used for performance indicators have similar meanings and the terms used vary within different industries. Here we will focus on leading and lagging indicators. The leading indicators are those that are typically associated with process input, whereas lagging indicators are most often associated with outputs.

10.2.2 PURPOSE AND USEFULNESS IN COMPLIANCE DETERMINATION

Leading and lagging performance indicators can be useful in a compliance program in a variety of ways. To reiterate the early discussion in Section 10.2, performance indicators are optimized if the criteria and metrics associated with them possess a strong correlation with the objectives, goals and targets of the performance indicators. The purpose of performance indicators is obviously to measure how a company or an organization is performing in relation to the goals and objectives that they have identified as being important. As previously discussed, these goals and objectives can be influenced or even defined by external expectations, internal factors and the corporate environmental strategy. The usefulness of performance indicators for environmental compliance is undeniable for any company in pursuit of a sustainable compliance program. Tracking and trending compliance-related activities can provide invaluable insight for corporate executives, environmental managers and risk managers that can help formulate strategies for such things as corrective actions, actions to prevent recurrences of noncompliance, process improvements and avoidance of regulatory violations, fines and penalties.

10.2.3 HOW TO USE LEADING INDICATORS

In general terms, leading indicators for environmental performance measure how well a company's processes and procedures protect against future compliance and environmental failures (Jacobs 2013). Leading indicators tend to be associated with process inputs, implying that they can be more easily influenced, but can be challenging to measure (van der Poel 2012). The fact that leading indicators are associated with process inputs means they are representative of in-process measures and reflective of how well a company's procedures and process implementation are performing (Global Environmental Management Initiative 1998). Examples of leading indicators include activities such as self-assessments, compliance audits, environmental training and employee surveys. A straightforward example of a leading indicator is the number of self-assessments conducted to evaluate compliance of a process or a set of regulated processes. Self-assessments are an input that can be influenced in a number of ways, including the frequency of assessments and the scope of the assessments. Processes that are experiencing issues with achieving or maintaining compliance can be assessed more frequently, and the scope of the audit can be tailored to address the specific details necessary to identify and correct process deficiencies. Figure 10.4 shows a simple graphical representation of the number of self-assessments versus identified compliance issues. In this case, the graph illustrates that with an increase in the number of self-assessments, there is a corresponding decrease in the number of identified compliance issues over time.

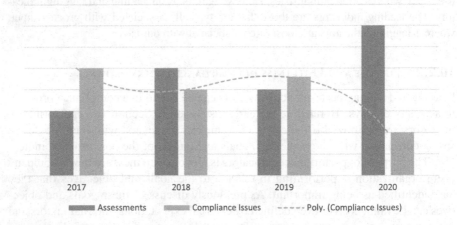

FIGURE 10.4   The number of assessments as a leading indicator.

Figure 10.4 represents a common situation for companies with regulatory obligations and environmental compliance responsibilities. Tracking how often they internally evaluate their operations and how compliance issues are being identified is an
effective leading indicator as to how well they are performing from a compliance
standpoint and is indicative of the types of and potential numbers of violations that
may result from an external audit by a regulatory agency.

Another straightforward example of a leading indicator is equipment maintenance
versus oil spills and releases. This leading indicator is often utilized by companies
with large fleets of heavy equipment to assess how maintenance schedules are contributing to releases of oil and fuel due to broken hydraulic lines, hoses, fuel lines,
etc. Not only do such occurrences contribute to equipment downtime, but there can
also be a significant amount of resources required in responding to and mitigating
releases of hydrocarbons and other contaminants into the environment. The graph in
Figure 10.5 illustrates equipment maintenance as a leading indicator for both equipment downtime and number of spills.

Training is another example of a leading indicator that can be used effectively
within a compliance program. Training elements such as classroom instruction
and required readings of relevant procedures, safety documents and compliance-
related protocols can directly translate to higher levels of worker awareness relative to compliance-related responsibilities and expectations. The commitment of a
company to support a training program with applicable and high-quality environmental content can have a positive influence on workers' level of understanding of
environmental compliance and regulatory issues that are related to their areas of
responsibility. Figure 10.6 illustrates a graphical representation of the relationship
between training as a leading indicator and compliance issues identified during
internal assessments and violations resulting from external compliance audits by a
regulatory agency.

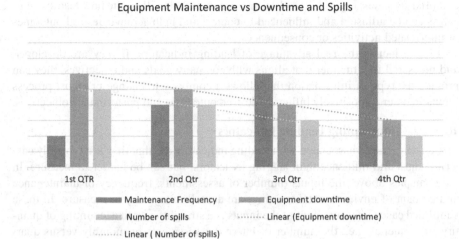

FIGURE 10.5 Equipment maintenance as a leading indicator.

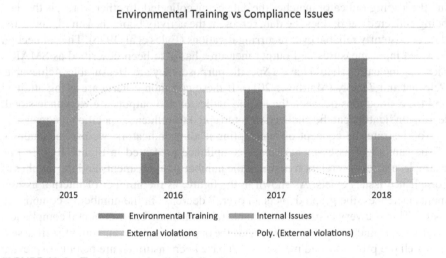

FIGURE 10.6 Training as a leading indicator.

In this example, there again is an obvious correlation between the leading indicator (number of environmental training classes) and the number of compliance issues identified during internal self-assessments as well as the number of violations identified by external regulatory agencies during compliance audits. The graph depicts a decrease in training from 2015 to 2016, which results in an increased number of compliance issues and regulatory violations. This is followed by an increase in training in 2017 and 2018, resulting in a decrease in both internal issues and regulatory violations.

These examples are simplified to illustrate how leading indicators are correlated to other factors that can impact company resources and potential environmental compliance issues. Figures 10.4–10.6 are also intended to illustrate how leading indicators can be adjusted and influenced to control and help achieve desired outcomes of interrelated activities or consequences.

This in fact is the real advantage of leading indicators. If they are developed and measured in a manner that aligns with company goals and objectives, they can provide the type of information that allows a company to change or adjust process inputs in order to positively influence desired outcomes and performance objectives.

10.2.3.1 Interrupting Leading Indicators

The characteristics associated with leading indicators are that they are typically easier to adjust and influence than lagging indicators, but can be harder to measure. In the examples above, the inputs (number of assessments, frequency of maintenance and amount of environmental training) are actually very easy to measure. In these simplified cases, measurement of the inputs is a straightforward accounting of quantity or frequency, i.e., the number of internal assessments or monthly versus quarterly maintenance. What commonly can present challenges with leading indicators is the ability to properly identify what outputs are affected by and associated with specific inputs or leading indicators. It is important that the original intent of selecting the leading indicator matches the data being collected. In other words, is the data being collected valid, does the data represent the activity being measured, and is the data consistently reliable over recurring iterations (Eyles et al. 1996). The connection between input measures and output measures has also been described as SMART, meaning measures should have a Specific purpose, they are Measurable, Achievable, Relevant and Timely (Matthews 2011). If these common criteria are being attained and the connections between the leading indicator and outputs are well understood, leading indicators can be changed or interrupted to influence specific outcomes.

To illustrate the concept of interrupting leading indicators, we use the example of the number of assessments versus compliance presented in Figure 10.4. In this example, a direct correlation between the number of assessments and the number of compliance issues exists. As shown in the figure, as the number of internal assessments increases, the graph depicts an overall decrease in the number of compliance issues. This is a very common observable relationship in environmental compliance. Internal compliance assessments provide the opportunity for companies to self-assess how well the protocols and measures that have been instituted are helping to prevent and mitigate noncompliant conditions. As the number of self-assessments increases, so do the opportunities for identifying and correcting operational or procedural issues, leading or contributing to noncompliant conditions. In this case, interrupting the leading indicator of self-assessments by increasing the number would theoretically result in fewer compliance issues over a period of time.

The relationship between the leading indicator (input) and the associated outcomes is obviously not always that simplistic and straightforward. There are many factors that can affect how directly the input is associated with the outcome of the performance aspect being measured. For example, if the personnel performing the assessments do not have adequate training or experience in conducting assessments, there could be deficiencies in the quality of the assessments and perhaps existing compliance issues are not being properly identified and continue to exist undetected. This would obviously be a problem associated with the training and skills of the personnel performing the assessments. This type of problem may be more difficult to measure and detect if personnel training and qualification are not an input consideration. This type of concern commonly shows up when internal self-assessments indicate low numbers of compliance issues but external regulatory audits find large numbers of compliance issues, as shown in Figure 10.7.

In Figure 10.7, there is a correlation between the number of self-assessments and the number of internally identified compliance issues. This information by itself would appear to indicate that increasing the number of self-assessments is positively affecting compliance by lowering the number of issues being identified internally. However, when considering the number of compliance issues identified by external regulators, the graph indicates that there are consistently more issues identified by the regulators than are being identified internally. In this case, the benefit of internally self-evaluating performance is not being realized and self-assessments are not effectively identifying and correcting issues or lowering company compliance risks. Figure 10.8 shows a graphical representation with the inclusion of personnel training and qualification regarding conducting assessments as an input and leading indicator in conjunction with number of assessments.

In this case, we see a positive correlation between the number of assessments, assessment training and the number of regulator-identified issues. In the example, the number of assessments and the assessment training both are considered leading indicators and can be adjusted individually or simultaneously to influence the number of compliance issues being identified internally as well as by external regulatory audits. This example also reiterates the earlier point regarding the importance of ensuring the leading indicators are measuring the intended objectives and that measurement data is reliable, representative and valid.

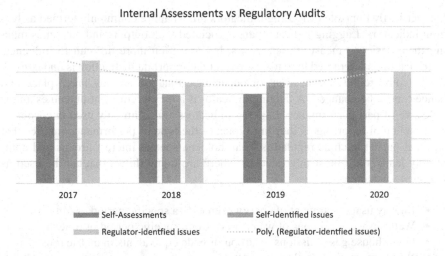

FIGURE 10.7 Self-assessments and internally identified issues versus issues identified by regulators.

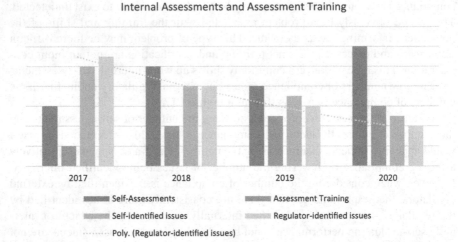

FIGURE 10.8 Self-assessments and assessment training and qualification as leading indicators.

In each of the examples above, there is an obvious relationship between the leading indicators that are the inputs and the associated outcomes. The ability to influence the outcomes by interrupting inputs (leading indicators) is a very powerful tool for managing performance, but of equal importance are outputs (lagging indicators), how they relate to leading indicators and how to effectively utilize them to measure and influence performance.

10.2.4 How to Use Lagging Indicators

Diametrically opposite of leading indicators are what are commonly termed as lagging indicators. Lagging indicators are associated with outputs and outcomes making them easier to measure than inputs, but are often more difficult to influence. Clarification is warranted here to ensure that an important distinction is understood. An outcome represents an event or occurrence that has already taken place and hence cannot be changed. A lagging indicator is a measurement of outcomes and is reflective of past performance. Lagging indicators are commonly used in corporate environmental reporting as they are typically the type of performance metrics that external parties such as regulators, shareholders, special interest groups and community leaders are most interested in. There are many things that can be used as lagging indicators:

- Energy usage – amount of energy used over a specific period of time
- Water usage – volumes of water used over a specific period of time
- Greenhouse gas emissions – carbon dioxide equivalents in metric tons
- Waste generation – volumes of waste generated for a specific time period
- Waste treatment – volume and types of waste treated for a given time period

- Number of spills – number of spills exceeding regulatory reporting threshold
- Wastewater discharges – volume of permitted wastewater discharges
- Remediation – progress of cleaning up contaminated sites (volumes of soil, water, etc.)
- Monitoring results – groundwater, air, vadose zone, storm water, etc.

While past performance itself cannot be altered, the lagging indicator being measured can provide important and invaluable insight into the occurrence of patterns in performance (Huether 2018).

10.2.4.1 Interrupting Lagging Indicators

Patterns in lagging indicators can in all actuality be a predictor of future performance. Oftentimes companies fail to realize the importance of past events as they are more focused on leading indicators because they are forward-looking and easier to influence. As alluded to in the previous discussion, leading indicators and lagging indicators can be very much interrelated from the perspective that every outcome is a result of an input or some combination of inputs. Interrupting a lagging indicator is not as straightforward as compared to leading indicators and requires an understanding of the inputs. Correlating the inputs to a specific outcome not only provides for the ability to measure past performance, but also the ability to influence future performance by correctly identifying and manipulating the input(s) which can directly affect outcomes.

Using waste generation as a lagging indicator, we can demonstrate how this interrelationship between lagging and leading indicators can be effective in enhancing performance. If Company ABC has set an objective of minimizing the volume of hazardous waste that they generate from their operations each year, their lagging indicator is simply the volume of waste that is generated on an annual basis. This is an easy-to-measure indicator that can be readily tracked and quantified. If this lagging indicator reveals that over the past three years, the volumes of hazardous waste generated has actually increased, Company ABC can use this trend in performance as an indication of the need to evaluate what causal factors might be leading to this pattern of increased volumes. Some potential causal factors for this example might include inadequate waste segregation practices (mixing nonhazardous wastes with hazardous wastes), using hazardous materials instead of comparable nonhazardous alternatives (product substitution), poor material handling practices that create excess waste and lack of or inadequate employee training. To graphically illustrate this relationship, Figure 10.9 compares the percent of product substitution to volumes of hazardous waste generated. The example depicts how an increase in product substitution affects the hazardous waste volume lagging indicator and how Company ABC could use a pattern in a lagging indicator to identify possible causes contributing to the pattern in an effort to effect change.

As with the previous examples in this chapter, we see the potential direct relationships between leading and lagging indicators. While Company ABC may not have identified product substitution as a leading indicator, this scenario demonstrates how it could be beneficial in interrupting an undesirable pattern of outcomes.

Product Substition vs Hazardous Waste Volumes

■ % Product substitution   ■ Volume of hazardous waste

FIGURE 10.9   Effect of product substitution on volumes of hazardous waste.

10.3   CONTINUOUS IMPROVEMENT

The overall goal of a continuous improvement program is to learn from previous activities and events; reduce the overall risk to company and employee; ensure compliance with all regulations and laws; and continue to provide quality products and services to its customers. A comprehensive improvement program strategically targets all elements of that program. According to Alston, Millikin and Piispanen (2018), the elements of a continuous improvement program are (1) define a performance baseline, (2) establish improvement goals and objectives, (3) identify and implement actions to meet goal and (4) measure and evaluate performance. This model has been revised and shown in Figure 10.10 to specifically address a continuous improvement for environmental compliance. Implementing an effective improvement program requires an improvement plan. A continuous improvement plan contains a set of targeted activities that are designed to produce gradual changes to an organization within a specific area through review, measurement and analyzing performance data. Without a plan, leaders and workers are not sure what to focus on or where improvements are needed or what is most important to the organization. The framework outlined in Figure 10.10 forms the basis for the continuous improvement plan. Each element should be explored thoroughly throughout the chain and once the end of the chain is reached and the information communicated to the entire workforce, the process begins again.

When an organization loses the propensity and desire to improve continuously the way business is being conducted, they have made the decision to stifle growth, productivity and sustainability. Continuous improvement should always be at the forefront of all business strategies, and workers should be encouraged by leadership to participate in the practice of always seeking better and more efficient ways of completing their tasks.

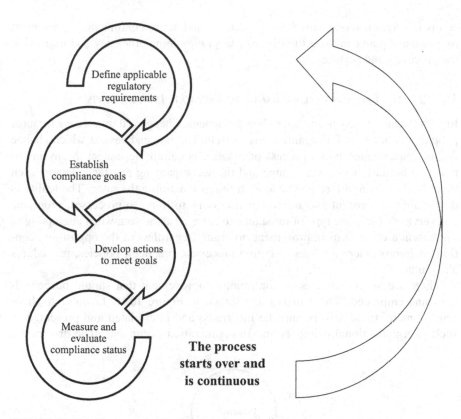

FIGURE 10.10 Environmental compliance continuous improvement chain.

10.3.1 USING KNOWLEDGE AND EXPERIENCE TO DRIVE IMPROVEMENTS

We often hear the adage that knowledge is power. This is true because knowledge provides the means to advance and empower people to achieve results. Knowledge is at the forefront of process improvement initiatives and has a significant role in a company's improvement strategy. Lasting knowledge can be gained through experience, direct exposure and education.

In order to learn from experiences effectively, it is necessary to reflect upon the experience that was encountered. This reflection can and should entail a detailed analysis of the event, causes, actions taken, preventive measures, positive and negative outcomes, etc.

Continuous improvement requires the effective use of specific information gained through knowledge and past experiences. The knowledge and experiences used may be obtained from internal personnel and experiences or from other organization groups or individuals. Benchmarking can be used to collect information and best practices from internal operations across a company or externally to collect and evaluate best practices within the industry. Benchmarking is a practice used by companies to compare their processes and practices to the best in the industry in which they compete. Organizations with a comprehensive and active continuous improvement program and plan can expect to discover ways of gaining efficiency and improving the processes and performance.

10.3.2 THE PROPENSITY FOR CHANGE TO FACILITATE IMPROVEMENTS

Improvement can only occur when change happens. Change is thus not easy for most people; it is not easy for organizations as well. The rate and ease at which change can be implemented into a process or practice is culture dependent. Many initiatives fail because it involves change and the accompanying resistance factors seen by individuals who are comfortable with things remaining the same. The reality is that change is constant and it is often necessary to improve processes, programs and services. The same type of resistance to change can be seen when attempting to implement a continuous improvement program. Regardless of the opposition, continuous improvement involves changing processes, practices, policies, procedures and culture.

There are several drivers of continuous improvement that should be considered and embraced. These drives are shown in Figure 10.11. In order to drive improvement, these drivers must be integrated and interrelated and presented as such to organizational members and in organization documents. The drivers are

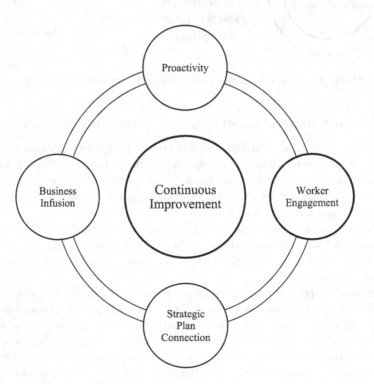

FIGURE 10.11   Continuous improvement drivers.

continuously considered, evaluated and their interrelation is infused in the business strategy and plan.

Proactivity: Stay on top of how work is performed and refrain from waiting until something fails or goes wrong before improving them. Being proactive demonstrates that there is a plan and propensity to improve.

Worker Engagement: Allow the workers to lead the way toward improvement since they have the most familiarity of ways to improve achievement of the work they perform. Therefore, it is smart business practice to get worker involvement and encourage them to use their knowledge and capability to assist leadership in driving improvement.

Strategic Plan Connection: Process improvement should be integrated into the strategic plan for a company and not implemented as a stand-alone process.

Business Infusion: Integration of process improvement in all aspects of the business to achieve optimal performance.

The continuous improvement drivers coupled with the desire for continuous improvement, a strategy implemented through a plan, the engagement of the leadership team and employees form the basis for having a solid continuous improvement culture that is effective and embraced by members of the organization that will facilitate improvements.

10.4 APPLIED LEARNING

Review the case study and respond to the questions that follow.

10.4.1 CASE STUDY

As the environmental compliance manager for a new company just starting out, you are responsible for defining the factors that are important to the company's environmental performance and developing methods to measure the performance of these factors so that they can be included in the annual environmental report. The company is a research and development company that performs a multitude of activities for various government and private clients. During the conduct of these activities, the company generates hazardous and municipal wastes, discharges wastewater, uses products that produce greenhouse gases and is responsible for monitoring and managing industrial storm water.

1. What aspects would you define as important for environmental performance?
2. Identify lagging indicators for each of these performance aspects?
3. Describe the metrics you would use for measuring these indicators.
4. What leading indicators might you use?
5. What actions would you take if a chosen performance area was not meeting expectations?
6. What should be considered in constructing a continuous improvement strategy and plan?

REFERENCES

Alston, Frances, Millikin, Emily J., & Piispanen, Willie. (2018). *Industrial hygiene improving worker health through an operational risk approach.* Boca Raton, FL: CRC Press/ Taylor & Francis.

Eyles, J., Cole, D., & Gibson, B. (1996). *Human health in ecosystem health: Issues of meaning and measurement.* Scientific, Windsor, ON: International Joint Commission (IJC) Digital Archive. Accessed April 20, 2020. https://scholar.uwindsor.ca/ijcarchive/503.

Eyles, John, & Furgal, Chris. (2002). Indicators in environmental health: Identifying and selecting common sets. *Canadian Journal of Public Health/Revue Canadienne De Sante'e Publique, 93,* S62–67. *JSTOR.* September/October. Accessed April 16, 2020 from www.jstor.org/stable/41993965.

Global Environmental Management Initiative. (1998). *Measuring environmental performance: A primer and survey of metrics in use.* Primer, Washington, DC: Global Environmental Management Initiative.

Hood, John, & Nicholl, Scott. (2002). The role of environmental risk management and reporting: An empirical analysis. *Journal of Environmental Assessment Policy and Management, 4*(1), 1–29. *JSTOR.* Accessed April 9, 2020 from www.jstor.org/stable/ enviassepolimana.

Huether, Derek. (2018. *Leading and lagging indicators.* February 7. Accessed April 20, 2020 from https://www.leadingagile.com/2018/02/leading-lagging-indicators/.

International Organization for Standardization. (2015). *Environmental management systems, third edition.* ISO 14001. Geneva: International Organization for Standardization, September 15.

Jacobs, Wendy B. (2013). *Suggested indicators of environmentally responsible performance of offshore oil and gas companies proposing to drill in the U.S. Arctic.* Paper, Cambridge, MA: Emmett Environmental Law & Policy Clinic, Harvard Law Clinic.

Kagan, Robert A., Thornton, Dorothy, & Gunningham, Neil. (2003). Explaining corporate environmental performance: How does regulation matter? *Law & Society Review, 37*(1), 51–90. *JSTOR.* March. Accessed April 13, 2020 from www.jstor.org/stable/1555070.

Matthews, Joseph R. (2011). Assessing organizational effectiveness: The role of performance measures. *The Library Quarterly: Information, Community, Policy, 81*(1), 83–110. doi: 10.1086/657447. *JSTOR.* January. Accessed April 20, 2020 from https://www.jstor.org/ stable/10.1086/657447.

National Research Council. (1999). *Industrial environmental performance metrics: Challenges and opportunities.* Washington, DC: The National Academies Press. Accessed April 17, 2020 from https://doi.org/10.17226/9458.

Scherpereel, Cécile, Van Koppen, C.S.A. (Kris), & Heering, G.B.F. (Frederique). (2001). Selecting environmental performance indicators: The case of numico. *Greener Management International, 33,* 97–114. *JSTOR.* Accessed April 9, 2020 from www. jstor.org/stable/greemanainte.33.97.

van der Poel, Karel. (2012). *KPILibrary.* Accessed April 17, 2020 from https://kpilibrary.c om/topics/lagging-and-leading-indicators.

Van Der Vorst, Rita, Grafe`-Buckens, Anne, & Sheate, William R. (1999). A systemic framework for environmental decision-making. *Journal of Environmental Assessment Policy and Management, 1,* 1–26. *JSTOR.* March. Accessed April 13, 2020 from www.jstor. org/stable/enviassepolimana.1.1.1.

Walczak, Magda. (2014). *What are key performance indicators (KPIs) and why you should use them.* August 25. Accessed April 16, 2020 from https://einsights.com/key-perf ormance-indicators-kpi/.

11 Lessons Learned

11.1 LEARNING FROM THE PAST TO PREVENT FUTURE OCCURRENCES

Learning from the experiences of the past is just what lessons learned are about. Lessons of the past can include lessons that resulted from the actions or activities that produced favorable or unfavorable conditions and results. Using the knowledge gained from performing activities to avoid the same or similar results is at the core of the purpose for utilizing lessons learned. Effective use of the concept involves having a credible process to identify, evaluate, document, validate and communicate what has been learned and its applicability to the business in which it is being applied. Most organizations have a lessons learned program that includes internal lessons learned as well as those learned from other industries or companies that have applicable business practices. In many cases, this program has assisted companies in averting unpleasant to catastrophic situations and improving processes and the quality of services and products provided to their customers. Figure 11.1 demonstrates the path that should be taken to utilize the lessons learned process effectively. Each element represented along the path is discussed below in A–E.

A. *Identify:* Identify the process or practices that can be exported and learned by employees, which can have positive or negative impacts on the business. The process or procedure should be the best in the industry or must help improve quality and increase productivity.

B. *Evaluate:* Evaluate carefully and completely the lessons learned and ensure that it is compatible with the intent of use. Ensure that the lessons learned are transportable to your business and can be repeated in a manner to gain optimal results as well as provide positive overall business impact.

C. *Document:* Annotate in detail the lessons learned and the process or procedure that is applicable for implementation. A detailed mapping of the process, program, procedure or project will be helpful especially for complex processes.

D. *Validate:* Ensure that the proposed lessons learned is in fact useful to the business or project and can be integrated into the business practices and processes for achieving the desired results.

E. *Communicate:* Communication of lessons learned should be timely and comprehensive and in writing, although verbal communication is also acceptable as long as it is followed by written means so that it can be retrieved as needed.

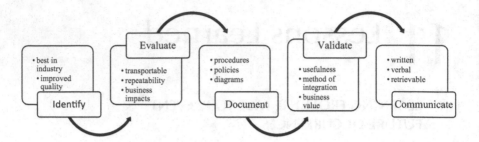

FIGURE 11.1   Lessons learned validation process.

Lessons learned can be positive and propelling or it can be negative and costly in various ways. However, the lessons learned that are frequently published and studied are those having negative impact, alerting others to beware so as to prevent the same event from happening in their organizations. Consider how much can be gained from publishing positive lessons learned in the same manner in which we see negative lessons learned documented and shared within companies and across industries. Figure 11.2 provides a pictorial view of the positive and negative lessons learned and the expected results.

Positive Lessons Learned: The impact of documenting and publishing positive lessons learned can help a company to replicate the activity, process or actions that can lead to a positive outcome in other areas of the company. This can help in improving product quality and customer relations, reduce injuries to people and the environment, reduce cost, facilitate process improvement and increase productivity. The knowledge gained through positive lessons learned when shared externally can help the industry improve across the globe.

Negative Lessons Learned: This type of lessons learned is typically what is published and shared to assist organizations in not repeating a process or practice that has had negative impact on process, equipment, people or the environment. Negative lessons learned often impact the process capability, people and the environment and are often costly to the organization in terms of downtime for workers and equipment to impacting a company's reputation. When identifying lessons learned, consider exploring answers to the following questions:

- What went well?
- What didn't work well?
- What could have been done differently?
- What were the impacts?
- Who or what was impacted?
- What events were not anticipated?

Lessons learned provide an opportunity to learn from the mistakes and woes of previous events or from the past events encountered by others. A comprehensive lessons learned program is instrumental to the continuous improvement process. Lessons learned can yield best practices for an organization or an industry when modeled and communicated properly.

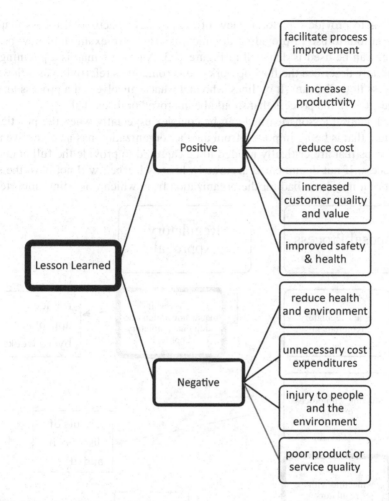

FIGURE 11.2   Lessons learned web chart.

11.2   CAPTURING AND DOCUMENTING LESSONS LEARNED

Organizations document procedure, policies and work practices to ensure that workers know what is expected of them and how to perform work correctly and safely and determine when process improvement is necessary. Proper documentation can provide management the ability to protect themselves when it comes to legal matters, audits, employee disputes and demonstrating compliance with rules and regulatory requirements. Documentation sets the stage on what is expected of everyone and how an organization operates and interacts with workers. Without documentation, organizations lack the record of how business is conducted and expectations for organizational members.

Reapplying lessons to prevent future mistakes is at the center of why it is necessary to capture lessons learned. A good method to use in capturing complex lessons learned is to provide a pictorial view of the process or activity. Process mapping is a great tool to use in providing this pictorial view. An example of how process mapping can be used is depicted in Figure 11.3. A process map is a planning tool that visually describes the flow of work and is sometimes referred to as a flowchart or process flow diagram. It outlines who and what is involved in a process and can divulge areas in the process that should be improved or discarded.

Documenting lessons learned can be complex especially when the practices or procedure that is being imported from another organization has an extensive number of steps that are critically needed to be captured to provide the full context of the process. If not documented accurately, lessons learned will not have the same impact as it may have had for the organization from which it is being imported. In

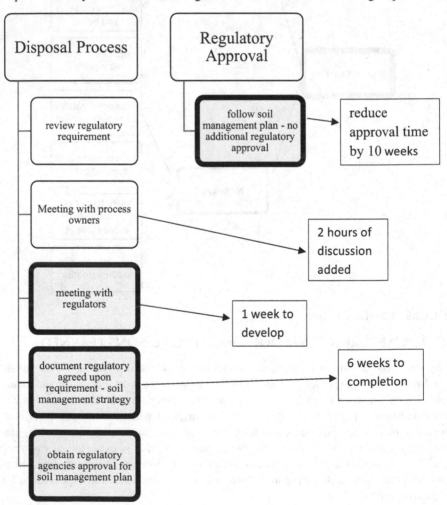

FIGURE 11.3 Lessons learned process maps soil management.

fact, it can have unintended impact and cause damage to the business, workers or products.

Figure 11.3 is a partial map of a process to gain regulatory approval for handling soil resulting from several large- and small-scale projects that took place on a Comprehensive Environmental Response Compensation and Liability Act (CERCLA) site, also known as a Superfund site. To ensure health and safety of the workers and the environment, regulatory compliance and project execution remain as scheduled, a soil management plan was developed with consultation with all of the applicable regulatory agencies' input in advance of project start. The plan was approved and signed off by all agencies providing approval to perform work and dispose of soil excavated based on the approved strategy. As long as work was performed in accordance with the strategy, there was no longer a need to gain regulatory approval for each project. This best practice resulted from the lessons learned from other CERCLA sites and the past experiences of having projects on hold while regulatory approval is sought and gained. This practice can cost a company their reputation in terms of following through with project commitment, loss of resources while waiting to move forward with work and the cost associated with paying workers that are not productive, equipment rentals, etc.

11.3 COMMUNICATING LESSONS LEARNED

The purpose of communication is exchange of information between humans. Communication is important in organizations because it forms the basis of how work is accomplished and directs organization life for workers. Leaders who are not skilled in communication tend to be viewed as ineffective and incompetent; thus, the organization will experience various levels of inefficiencies. The use of communication is not limited to the use of spoken words, but it also involves nonverbal and written words. Frequent, open and honest communication with employees is important for organizations to function effectively. Appropriate communication is key in ensuring success of transporting and implementing lessons learned, especially those that are being imported from outside of the organization. In fact, when considering changes gleaned from external entities, realize that the level of complexities may be increased because of the lessons learned that yielded a best practice is not from within the company. In such cases, the culture of the organizations will evidently be quite different, thus posing additional challenges.

When communicating information that is expected to be retained by workers, the communication methods that would represent the most value to leadership in a work setting is written and visual. Because the rate and mode of retention is different for many workers, a combination of all communication methods – verbal, nonverbal or written – should be considered and used to gain the maximum impact of learning and retention. When using the four communication channels, it is necessary to consider the use of the four paradigms listed in Figure 11.4 to achieve ultimate effectiveness.

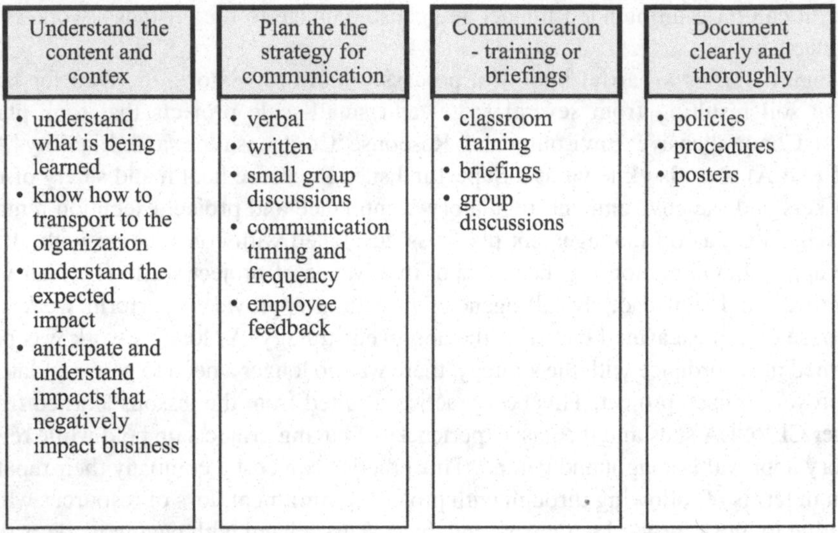

Understand the content and contex	Plan the the strategy for communication	Communication - training or briefings	Document clearly and thoroughly
• understand what is being learned • know how to transport to the organization • understand the expected impact • anticipate and understand impacts that negatively impact business	• verbal • written • small group discussions • communication timing and frequency • employee feedback	• classroom training • briefings • group discussions	• policies • procedures • posters

FIGURE 11.4   Communication paradigm.

In addition to the tips listed in Figure 11.4, additional considerations for communicating lessons learned to employees are:

- Inform employees early and completely. Inform workers how their work practices, policies and procedures will change. Communicate the impact to employees.
- Request feedback from employees.
- Develop and provide training if warranted. If training is not warranted, at least a briefing should be developed and provided to employees to ensure they understand the changes and work impact.

11.4   INTEGRATING LESSONS LEARNED TO PRODUCE CHANGE

Implementing lessons learned can and will fail if the organization is not prepared for the change. If the culture is not prepared, then the employees are not prepared. A culture that readily resists change is one that is difficult to meet the need of the organization for the future because the one constant in an organization is that change will happen. The culture of an organization provides significant clues as to how members of an organization will act and react when they are asked to change. Change management is not easy and can present added stress on the leadership team, employees and potentially on customers. Some actions that can be taken to assist with change management include communicate completely and early the reason why the change is necessary, know the culture of the organization and what it can tolerate; demonstrate leadership support and engagement; identify and solicit workers that are in support of the change and use them and change agents; and solicit feedback and buy-in from those who will be impacted.

Although change is not easy, management must prepare the organization for the inclusion of the changes that result from a lesson learned. Implementation of new processes systems, practices or policies is never easy, although it is necessary. The same is true with lessons learned implementation. When implemented appropriately and the new process suits the needs of the organizations, there is a great benefit to the employees as well as the organization. Some of the yields will be noted across the entire process.

11.5 APPLIED LEARNING

The case study below was outlined by Alston in the book entitled *Lean Implementation Applications and Hidden Costs 2017* to demonstrate implementation of a Lean project. This project will be used to elaborate on the value of lessons learned: method for transporting and communicating lessons learned. Review the case study and respond to the following questions:

1. What lessons learned can be transported?
2. List and discuss three important elements of the communication plan that should be implemented.
3. Discuss how the communication paradigm pictured in Figure 11.4 can be used in implementing the lessons learned gleaned from the case study.
4. How was process mapping used in implementing the lessons learned? Was it deemed to be effective?

11.5.1 CASE STUDY

This case study involves identifying and implementing a process to track chemicals from cradle to grave. The lessons learned involved implementation of process coupled with technology to track chemicals from the point of purchasing. The previous process used for inventory chemicals required technicians to physically handle each chemical, which subjected workers to hazards that arose from handling a variety of chemicals in varying sized, weight, container configuration and hazardous content. Also, a major lesson learned from utilizing the hands-on practice of conducting the inventory created ergonomic related challenges for workers yielding injuries for several workers. Completing the inventory required workers to lift heavy containers and place their body in stressful positions.

The current process was mapped to ensure a complete understanding of the 'as is' state. Process mapping is one of the simplest techniques used for identifying inefficiencies and waste and streamlining process components. To ensure a complete understanding of the issue and integration of the changes in process, workers were interviewed, a job hazard analysis was reviewed and revised where necessary and work observations were performed. The revised process was processed mapped with engagement of workers. The process mapping of current (as is) process as well as the revised process (to be) is showed in Figures 11.5 and 11.6 (Alston 2017).

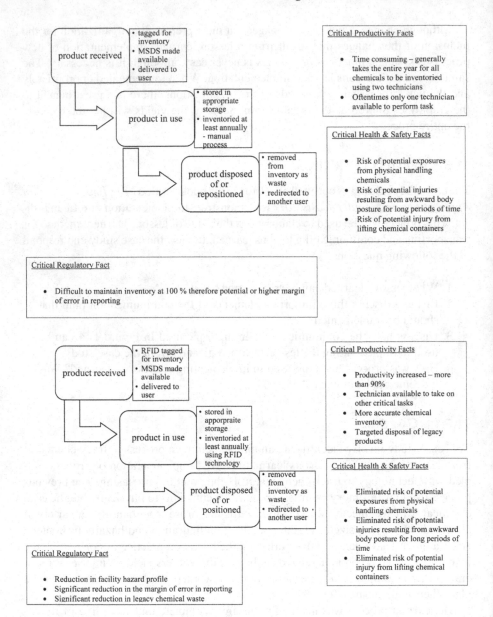

FIGURE 11.5 'As is' process.

Note: Additional information on the case study can be found in chapter 10 of the reference book (Alston 2017). Reading this chapter should provide additional information and expand the considerations for answering the questions on lessons learned.

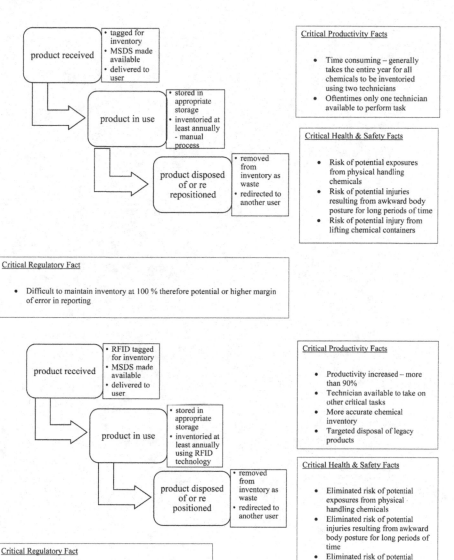

FIGURE 11.6   'To be' process.

REFERENCE

Alston, Frances. (2017). *Lean implementation applications and hidden costs*. Boca Raton, FL: CRC Press.

12 Compliance and Implementation Toolkit

This chapter is designed to assist users in designing checklists that can be used to evaluate and assess the performance of their environmental programs and ultimate compliance. Environmental compliance requires strategic planning and evaluation that is continuous and changes as processes and practices change and as areas of improvement are discovered. The compliance and implementation toolkit contains a series of checklists and tools intended for guidance and instructional purposes. The items within the toolkit were adapted from checklists and guidance developed by the Environmental Protection Agency (EPA), federal agencies and state regulatory agencies that are available in the public domain. The checklists included in the toolkit are presented to assist users in assessing compliance with some of the common regulatory requirements, as presented in the regulatory sections of Chapter 3 in this book. The checklists are intended to provide a series of templates that users can modify to assist in evaluating compliance. The checklists provided in the appendices of this book were adapted from documents that were developed by the various aforementioned agencies over several years. The checklists and any modifications should be verified against current regulations at the federal, state and local levels as they apply to individual regulatory environments within each individual company. The checklist covering specified regulatory requirements are enlisted in Table 12.1.

TABLE 12.1
Regulatory Compliance Tools

Compliance Checklist	Appendix
RCRA	A
NPDES	B
CAA	C
EPCRA	D
SDWA	E
SPCC	F
FIFRA	G
TSCA	H
NEPA	I
CERCLA	J

APPENDIX A
Toolkit for Evaluating Resource Conservation and Recovery Act (RCRA)

This appendix provides a series of checklists intended to assist compliance managers and personnel performing inspections and assessments of some of the more common operations involving nonhazardous and hazardous wastes covered under RCRA. These tools are provided for instructional purposes and guidance only and are not intended to represent an all-inclusive collection of RCRA regulatory requirements. The tools provided in this appendix are adapted from EPA checklists and organized within this appendix as follows.

Nonhazardous Solid Waste (RCRA Subtitle D):

1. Storage and collection of solid wastes
2. Recycling

Hazardous Waste (RCRA Subtitle C):

1. Generator category determination
2. Waste stream characterization and determination
3. Very small quantity generator requirements
4. Small quantity generator requirements
5. Large quantity generator requirements
6. Used oil
7. Universal waste
8. Land disposal restrictions for soil and debris
9. Organic air emissions
10. Underground storage tanks

TABLE A.1
Storage/Collection of Solid Waste Checklist (applies to residential, commercial and institutional solid wastes and street wastes)

Date: Time:

Inspector's name: Signature:

1. Are all solid wastes stored so as not to cause a fire, health or safety hazard?   Yes___ No ___

2. Are all solid wastes containing food wastes stored securely in covered or closed containers which are nonabsorbent, leakproof, durable, easily cleaned if reused and designed for safe handling?   Yes___ No___

3. Are solid waste containers of an adequate size and number contain all food wastes, rubbish and ashes generated between collections?   Yes___ No___

4. Are bulky wastes stored so as not to create a nuisance and to avoid the accumulation of solid wastes and water in and around the bulky items?   Yes___ No___

5. Are reusable containers that are emptied manually capable of being serviced without the collector coming into contact with the waste?   Yes___ No___

6. Do waste containers used for the storage of solid wastes (or materials which have been separated for recycling) meet the standards established by ANSI for waste containers as follows: Waste Containers – Safety Requirements, ANSI Z245.30-1994; Waste Containers – Compatibility Dimensions, ANSI Z245.60-1996?   Yes___ No___

7. Are solid wastes which contain food wastes collected at a minimum of once during each week?   Yes___ No___

8. Are bulky wastes collected at a minimum of once every three months?   Yes___ No___

9. Are all wastes collected with sufficient frequency to inhibit the propagation or attraction of vectors and the creation of nuisances?   Yes___ No___

TABLE A.2
Solid Waste Recycling Checklist

Date: Time:

Inspector's name: Signature:

1. Does a solid waste reduction program exist?	Yes___ No___
2. Is the recycling program in compliance with applicable state/local requirements?	Yes___ No___
3. Are reusable or marketable materials collected at regular intervals?	Yes___ No___
4. Are there over 100 office workers?	Yes___ No___
5. Are high-grade papers separated at the source of generation?	Yes___ No___
6. Are high-grade papers separately collected?	Yes___ No___
7. Are high-grade papers sold for recycling?	Yes___ No___
8. Does the facility generate 10 tons or more of waste corrugated containers per month?	Yes___ No___
9. Are the waste corrugated containers collected separately?	Yes___ No___
10. Are the waste corrugated containers sold for the purposes of recycling?	Yes___ No___

Describe any issues identified during the inspection and corrective actions taken:

TABLE A.3
Generator Category Determination Date: _____

Generators of hazardous waste are required to determine their generator category based on the amount of hazardous waste generated from month to month in each separate category below and select the most stringent generator category in Table A-1.

Acute hazardous waste

1. Are acute hazardous wastes generated at the facility?   Yes___ No___
2. Amount of acute hazardous waste generated per month?   _____kg

Non-acute hazardous waste

1. Are non-acute hazardous wastes generated at the facility?   Yes___ No___
2. Amount of non-acute hazardous waste generated per month?   _____kg

Residues from cleanup of acute hazardous waste

1. Are residues from acute hazardous waste cleanup generated at facility   Yes___ No___
1. Amount of residues from acute hazardous waste cleanup generated per month   _____kg

Quantity of acute hazardous waste generated in a calendar month	Quantity of non-acute hazardous waste generated in a calendar month	Quantity of residues from a cleanup of acute hazardous waste generated in a calendar month	Generator category
>1 kg	Any amount	Any amount	Large quantity generator
Any amount	≥1,000 kg	Any amount	Large quantity generator
Any amount	Any amount	>100 kg	Large quantity generator
≤1 kg	>100 kg and <1,000 kg	≤100 kg	Small quantity generator
≤1 kg	≤100 kg	≤100 kg	Very small quantity generator

TABLE A.3
Generator Categories (*Source:* 40 CFR 262.13)

The following hazardous wastes are not included when making the monthly quantity-based determinations:

1. Waste which is exempt from regulation under 40 CFR 261.4(c) through (f), 261.6(a)(3), 261.7(a) (1) or 261.8
2. Waste that is managed immediately upon generation only in on-site elementary neutralization units, wastewater treatment units or totally enclosed treatment facilities as defined in 40 CFR 260.10
3. Waste that is recycled, without prior storage or accumulation, only in an on-site process subject to regulation under 40 CFR 261.6(c)(2)
4. Used oil managed under the requirements of 40 CFR 261.6(a)(4) and 40 CFR part 279
5. Spent lead-acid batteries managed under the requirements of 40 CFR part 266 subpart G
6. Universal waste managed under 40 CFR 261.9 and 40 CFR part 273
7. Hazardous waste that is an unused commercial chemical product (listed in 40 CFR part 261 subpart D or exhibiting one or more characteristics in 40 CFR part 261 subpart C) that is generated solely as a result of a laboratory cleanout conducted at an eligible academic entity pursuant to §262.213
8. Is managed as part of an episodic event in compliance with the conditions of subpart L of this part.
9. Hazardous waste pharmaceutical, as defined in §266.500, that is subject to or managed in accordance with 40 CFR part 266 subpart P or is a hazardous waste pharmaceutical that is also a Drug Enforcement Administration–controlled substance and is conditionally exempt under §266.506
10. Hazardous waste when it is removed from on-site accumulation, so long as the hazardous waste was previously counted once
11. Hazardous waste generated by on-site treatment (including reclamation) of the generator's hazardous waste, so long as the hazardous waste that is treated was previously counted once
12. Hazardous waste spent materials that are generated, reclaimed and subsequently reused on-site, so long as such spent materials have been previously counted once

TABLE A.4

Waste Stream Characterization Checklist

1. Describe the process generating the waste.
2. Rate of generation (i.e., volume per week, month or year).
3. Describe the physical and chemical properties of wastes:
4. Waste is characterized by:

   Process knowledge   ☐

   Waste analysis   ☐ Date(s) sampled _____
5. Does the waste meet the definition of a solid waste (40 CFR 261.2)?   ☐ Yes ☐ No
6. Is the waste excluded or exempt from hazardous waste regulations (40 CFR 261.4)?   ☐ Yes ☐ No
7. Is the waste a listed hazardous waste?

   F-listed (40 CFR 261.31)   ☐ Yes ☐ No

   Waste codes: _____

   K-listed (40 CFR 261.32)   ☐ Yes ☐ No

   Waste codes: _____

   P-listed (40 CFR 261.33(e)) ☐ Yes ☐ No

   Waste codes: _____

   U-listed (40 CFR 261.33(f)) ☐ Yes ☐ No

   Waste codes: _____
8. Is the waste a characteristic hazardous waste? ☐ Yes ☐ No

   Ignitable (D001) ☐ Corrosive (D002) ☐ Reactive (D003) ☐

   Toxic ☐ waste codes: _____
9. Waste stream determination

   Nonhazardous ☐

   Hazardous ☐ Waste codes: _____ _____
10. Do Land Disposal Restrictions (LDRs) apply to waste stream? ☐ Yes ☐ No
11. Identify any Underlying Hazardous Constituents (UHCs) (40 CFR 268.40 (e))

TABLE A-5
Very Small Quantity Generator Hazardous Waste Checklist

Date: Time:

Inspector's name: Signature:

Records and Documentation

1. Facility continues to meet Very Small Quantity Generator category <100 kg of hazardous waste and <1 kg of P-listed waste generated per month?   Yes___ No___

2. Are waste hazardous waste determinations made at the point of generation before any dilution, mixing or other alteration of the waste occurs, and at any time in the course of its management where properties may have changed 262.14(a) (2)→262.11(a)?   Yes___ No___

3. Determinations made as to whether waste meets any of the listings in 40 CFR 261 Subpart D-262.14(a)(2)→262.11(c)?   Yes___ No___

4. Determinations made as to whether waste exhibits any of the characteristics identified in 40 CFR 261 Subpart C-262.14(a)(2)→262.11(d)?   Yes___ No___

5. On-site accumulation of waste is <1,000 kg at all times?   Yes___ No___

6. For waste consolidated at Large Quantity Generator (LQG) site under the control of the same person as this VSQG facility are the containers marked as 'Hazardous Waste' 262.14(a)(5)(viii)(B)(1)?   Yes___ No___

7. Are these same containers labeled with an indication of the nature of the hazard-262.14(a)(5)(viii)(B)(2)?   Yes___ No___

8. Is waste sent off-site to a permitted TSDF, a recognized municipal solid waste landfill, a recognized non-municipal nonhazardous waste landfill, a legitimate recycler or an LQG under control of the same person as the VSQG-262.14(a)(5)?   Yes___ No___

TABLE A.6
Small Quantity Generator Hazardous Waste Checklist

Date: Time:

Inspector's name: Signature:

Records and Documentation

9. Facility continues to meet the Small Quantity Generator category >100 kg and
   <1,000 kg of hazardous waste and <1 kg of P-listed waste generated per month?   Yes___ No___

10. Waste is accumulated on-site for ≤180 days or ≤270 days if transporting greater
    than 200 miles if a non-permitted TSDF?   Yes___ No___

11. On-site accumulation of waste is <6,000 kg at all times?   Yes___ No___

12. Manifest system is used per 40 CFR 262.20(a)(1)?   Yes___ No___

13. Manifests are maintained for three years? (40 CFR 262.40(a))   Yes___ No___

14. Has obtained an EPA I.D. number? (40 CFR 262.20(a))   Yes___ No___

15. Manifests contain generator name, address, phone number? (262.20(a))   Yes___ No___

16. Transporter(s) name & EPA I.D. number recorded on manifest? (40 CFR
    262.20(a))   Yes___ No___

17. Designated facility name, address and EPA I.D. number included on manifest?
    (40 CFR 262.20(a))   Yes___ No___

18. Alternative facility designated on manifest (optional)? (40 CFR 262.20(c))   Yes___ No___

19. Has unique preprinted manifest tracking number and number of pages on
    manifest? (40 CFR 262.20(a))   Yes___ No___

20. DOT shipping name, hazard class, waste code and reportable quantity (RQ) (if
    required by 49 CFR 172) included on manifest? (40 CFR 262.20(a))   Yes___ No___

21. Number, type, quantity, unit wt./vol. of containers recorded on manifest? (40 CFR
    262.20(a))   Yes___ No___

22. Proper certification including waste minimization recorded on manifest? (40 CFR
    262.20(a))   Yes___ No___

23. Manifests are signed and dated by generator and transporter? (40 CFR 262.23(a)) Yes___ No___

24. Exception reports have been submitted if necessary? (40 CFR 262.42)   Yes___ No___

25. LDR notification/certification are sent with manifests on first shipment?
    (40 CFR 262.16(b)(7)→268.7(a)(2))   Yes___ No___

26. LDR notification/certification includes manifest number, correct EPA
    waste codes and treatment standards and waste analysis data? (40 CFR 262.
    16(b)(7)→268.7(a)(2))   Yes___ No___

27. LDR notification/certification/waste analysis data and other documents are
    maintained for three years? (40 CFR 262.16(b)(7)→268.7(a)(8))   Yes___ No___

TABLE A.7
Small Quantity Generator Hazardous Waste Checklist

Date: Time:

Inspector's name: Signature:

Waste Analysis/Waste Determination and Land Disposal Restrictions

1. Waste is not diluted impermissibly to meet LDR standards? (40 CFR262.16(b)(7)→268.3(a) & (b))

2. Are hazardous waste determinations made at the point of generation before any dilution, mixing or other alteration of the waste occurs and at any time in the course of its management where properties may have changed? (262.14(a)(2)→262.11(a))   Yes___ No___

3. Determinations made as to whether waste meets any of the listings in 40 CFR 261 Subpart D? (262.14(a)(2)→262.11(c))   Yes___ No___

4. Determinations made as to whether waste exhibits any of the characteristics identified in 40 CFR 261 Subpart C? (262.14(a)(2)→262.11(d))   Yes___ No___

5. All applicable EPA hazardous waste numbers are identified? (40 CFR 262.11(g))   Yes___ No___

6. Records supporting hazardous waste determinations are maintained for at least three years from the date that the waste was last sent to on-site or off-site treatment, storage or disposal? (40 CFR 262.11(f))   Yes___ No___

7. Waste determinations, including results of any tests, sampling, waste analyses or other determinations; tests, sampling and analytical methods; descriptions or processes, waste composition and waste properties; and records which explain the knowledge basis are documented? (40 CFR 262.11(f))   Yes___ No___

8. Are determinations made for waste that does not meet applicable treatment standards (ATS)? (40 CFR262.16(b)(7)→268.7(a)(2))   Yes___ No___

9. One-time written notice to TSDF has been submitted with initial shipment and a copy placed in file? (40 CFR 262.16(b)(7)→268.7(a)(2))   Yes___ No___

10. Is there waste covered by a National Capacity Variance(s)-268 Subpart C, Extension or Petition? (40 CFR 262.16(b)(7)→268.5 & 6)   Yes___ No___
    Describe the variance, extension or petition that applies

11. For waste shipped off-site for disposal, has a notice been provided to the land disposal facility with the initial shipment or a revised notice if changes occur, stating that the waste is exempt from the LDRs? (40 CFR 262.16(b)(7)→268.7(a)(4))   Yes___ No___

12. For waste shipped off-site for disposal, has a notice been provided with the initial shipment or new notification, if changes occur? (40 CFR 262.16(b)(7)→268.7(a)(2))   Yes___ No___

13. For waste shipped off-site for disposal, are the following included on the LDR notice: EPA hazardous waste number(s), manifest number(s), waste analysis data, if available, and waste constituents; wastewater or non-wastewater classification; and subcategory, if applicable? (40 CFR 262.16(b)(7)→268.7(a)(2) and (a)(4))   Yes___ No___

14. If waste is determined to be excluded from the definition of hazardous or solid waste, is a one-time notice describing generation, subsequent exclusion or exemption and disposition of the waste retained in the facility's on-site files? (40 CFR 262.16(b)(7)→268.7(a)(7))   Yes___ No___

15. If generator determines waste or soil contaminated with waste does meet the ATS or does not exceed prohibition levels and requires no further treatment, has a one-time written notice been submitted to the TSDF with initial shipment and a copy placed in file? (40 CFR 262.16(b) (7)→268.7(a)(3)(i))   Yes___ No___

16. Have underlying hazardous constituents as defined in 40 CFR 268.2(i) been determined for waste that is not D001 non-wastewater? (40 CFR 262.16(b)(7)→268.9(a))   Yes___ No___

17. For waste that is land disposed, has it been determined if waste meets the treatment standards specified in 268 Subpart D? (40 CFR 262.16(b)(7)→268.9(c))   Yes___ No___

18. For claims that a characteristic waste is no longer hazardous, has a one-time notification and certification been sent to EPA or authorized state, a copy placed in the file and updated if there are changes in process, operation or receiving facility? (40 CFR 262.16(b)(7)→268.9(d))

                                                                                               Yes___ No___

TABLE A.8
Hazardous Waste Satellite Accumulation Area Checklist

Date: _____Time:_____

Inspector's name: _____Signature:_____

1. Is SAA at or near the point of generation?--☐ Yes ☐ No
2. IS SAA under control of the responsible operator? --- ☐ Yes ☐ No
3. Is quantity of waste less than 55 gals (or 1 kg/1 quart of P-listed waste)? -------------- ☐ Yes ☐ No
4. Waste exceeding above quantities in SAA transferred within three days? -------------- ☐ Yes ☐ No
5. Containers labeled with words 'Hazardous Waste' (262.15(a)(5)(i))?-------------------☐ Yes ☐ No
6. Containers labeled with nature of the hazard (262.15(a)(5)(ii))?------------------------ ☐ Yes ☐ No
7. Are containers kept closed when not adding waste? --------------------------------------- ☐ Yes ☐ No
8. Are containers in good condition? --☐ Yes ☐ No
9. Are containers compatible with waste being stored?---------------------------------------☐ Yes ☐ No
10. Incompatible waste not being placed into the same container?---------------------------☐ Yes ☐ No
11. Wastes are not placed in an unwashed container that previously held an
    incompatible waste?- --☐ Yes ☐ No
12. Containers of incompatible waste are kept separated?-------------------------------------☐ Yes ☐ No

Describe any issues identified during the inspection and corrective actions taken:

SAA ID:_____ SAA location: _____

Person responsible for SAA:_____

Description of problems:_____

Actions taken: _____

SAA ID:_____ SAA location: _____

Person responsible for SAA:_____

Description of problems:_____

Actions taken: _____

SAA ID:_____ SAA location: _____

Person responsible for SAA:_____

Description of problems:_____

Actions taken: _____

TABLE A.9
Small Quantity Generator Hazardous Waste Checklist

Date: _____ Time: _____

Inspector's name: _____ Signature: _____

<u>Container Accumulation Areas (CAAs)</u>

1. Containers are in good condition? (40 CFR 262.16(b)(2)(i))   Yes___ No___
2. Containers used are compatible with waste? (40 CFR 262.16(b)(2)(ii))   Yes___ No___
3. Containers are kept closed? (40 CFR 262.16(b)(2)(iii)(A))   Yes___ No___
4. Containers are not opened, handled or managed in a manner to cause them to
   leak? (40 CFR 262.16(b)(2)(iii)(B))   Yes___ No___
5. CAA is inspected weekly? (40 CFR 262.16(b)(2)(iv))   Yes___ No___
6. Incompatible wastes are not placed in the same container? (40 CFR 262.16(b)
   (2)(v)(A))   Yes___ No___
7. Wastes are not placed in an unwashed container that previously held an
   incompatible waste? (40 CFR 262.16(b)(2)(v)(B))   Yes___ No___
8. Containers of incompatible waste are separated or otherwise protected from each
   other? (40 CFR 262.16(b)(2)(v)(C))   Yes___ No___
9. Containers are labeled as 'Hazardous Waste'? (40 CFR 262.16(b)(6)(i)(A))   Yes___ No___
10. Containers are labeled with an indication of the nature of the hazard? (40 CFR
    262.16(b)(6)(i)(B))   Yes___ No___
11. Containers are marked with accumulation start dates? (40 CFR 262.16(b)(6)(i)(C)) Yes___ No___

PRE-TRANSPORT REQUIREMENTS

12. Containers are packed, labeled and marked per DOT requirements? (40 CFR
    262.30, 262.31, 262.32, respectively)   Yes___ No___
13. Placards are provided for use by transporters when applicable? (40 CFR 262.33) Yes___ No___
14. Containers are marked with all applicable EPA hazardous waste numbers? (40
    CFR 262.11(g))   Yes___ No___

TABLE A.10

Small Quantity Generator Hazardous Waste Checklist

Date: _____ Time: _____

Inspector's name: _____ Signature: _____

Preparedness/Prevention and Training

1. An emergency coordinator has been designated and that person is on-site or
   on-call? (40 CFR 262.16(b)(9)(i))   Yes___ No___

2. Emergency coordinator's name and phone number, fire department's phone
   number and locations of fire extinguishers and spill control equipment has been
   posted near the phone? (40 CFR 262.16(b)(9)(ii))   Yes___ No___

3. Arrangements have been made with local emergency agencies? (40 CFR
   262.16(b)(8)(vi)(A))   Yes___ No___

4. Coordinating agencies have been familiarized with layout, waste types, access
   points, evacuation routes and likely casualty types? (40 CFR 262.16(b)(8)(vi)(A)
   (2))   Yes___ No___

5. A primary agency has been designated in the event of more than one responding
   FD or PD? (40 CFR 262.16(b)(8)(vi)(A)(3))   Yes___ No___

6. Records documenting arrangements with response agencies are maintained? (40
   CFR 262.16(b)(8)(vi)(B))   Yes___ No___

7. Immediately responds to or coordinates with a response agency or contractor in
   the event of a fire or spill? (40 CFR 262.16(b)(9)(iv))   Yes___ No___

8. All employees have been familiarized with waste handling and emergency
   procedures relevant to their responsibilities? (40 CFR 262.16(b)(9)(iii)) Yes___ No___

TABLE A.11
Small Quantity Generator Hazardous Waste Checklist

Date: _____ Time: _____

Inspector's name: _____ Signature: _____

Required Response Equipment and Hazard Management

Note: 40 CFR 262.15(a)(7) requires this equipment be available even if waste is accumulated only in SAAs.

1. Facility is operated to minimize possibility of a fire, explosion or release? (40 CFR 262.16(b)(8)(i))   Yes___ No___

2. An internal communications or alarm system is available and operational? (40 CFR 262.16(b)(8)(ii)(A))   Yes___ No___

3. Devices are available for and capable of summoning emergency assistance? (40 CFR 262.16(b)(8)(ii)(B))   Yes___ No___

4. Adequate supply and proper spill control, decontamination and safety equipment (fire blankets, respirators, absorbent, etc.) is available? (40 CFR 262.16(b)(8)(ii)(C))   Yes___ No___

5. An adequate water supply for fire control equipment is available? (40 CFR 262.16(b)(8)(ii)(D))   Yes___ No___

6. Communication and emergency equipment are tested and maintained? (40 CFR 262.16(b)(8)(iii))   Yes___ No___

7. Communication equipment is immediately accessible when waste is being handled? (40 CFR 262.16(b)(8)(iv)(A))   Yes___ No___

8. Communication equipment is immediately accessible when only one person present on-site? (40 CFR 262.16(b)(8)(iv)(B))   Yes___ No___

9. Adequate aisle space is provided for type of waste management and emergency equipment used? (40 CFR 262.16(b)(8)(v))   Yes___ No___

TABLE A.12
Small Quantity Generator Hazardous Waste Checklist

Date: _____ Time:_____

Inspector's name: _____ Signature: _____

Tank Accumulation Area(s)

1. Reactive waste and mixture of incompatible wastes are accumulated so that it does not: (1) generate extreme heat or pressure, fire or explosion or violent reaction; (2) produce uncontrolled toxic mists, fumes, dusts or gases in sufficient quantities to threaten human health; (3) produce uncontrolled flammable fumes or gases in sufficient quantities to pose a risk of fire or explosions; (4) damage the structural integrity of the device or facility containing the waste; (5) via other like means threaten human health or the environment? (40 CFR 262.16(b)(3)(ii) (A)→265.17(b)(1) through (b)(5))   Yes___ No___

2. Wastes are evaluated to ensure they are not likely to cause corrosion or other tank failure? (40 CFR 262.16(b)(3)(ii)(B))   Yes___ No___

3. Tanks have 2 feet of freeboard or containment system? (40 CFR 262.16(b)(3)(ii) (C))   Yes___ No___

4. Continuously fed tanks are equipped with feed cutoff system or bypass system? (40 CFR 262.16(b)(3)(ii)(D))   Yes___ No___

5. Waste feed cutoff/bypass system, monitoring equipment, freeboard and tank levels are inspected daily? (40 CFR 262.16(b)(3)(iii)(A))   Yes___ No___

6. Data gathered from pressure and temperature gauges are recorded daily? (40 CFR 262.16(b)(3)(iii)(B))   Yes___ No___

7. Waste levels are measured daily? (40 CFR 262.16(b)(3)(iii)(C))   Yes___ No___

8. Tanks with full secondary containment *and* documented leak detection equipment/ procedures are inspected weekly for items 3–5 above? (40 CFR 262.15 (b)(3)(iv))   Yes___ No___

9. Weekly inspections of tanks are conducted for leaks, discharges and corrosion? (40 CFR 262.16(b)(3)(iii)(D))   Yes___ No___

10. Weekly inspections of areas around tanks are conducted for leaks or discharges? (40 CFR 262.16(b)(3)(iii)(E))   Yes___ No___

11. Ignitable/reactive wastes are treated or accumulated so as to prevent ignition *or* uses a tank only for emergencies? (40 CFR 262.16(b)(3)(vii)(A))   Yes___ No___

12. Ignitable/reactive waste are treated or accumulated per NFPA's buffer zone requirements? (40 CFR 262.16(b)(3)(vii)(B))   Yes___ No___

13. Incompatible wastes are not placed in the same tank? (40 CFR 262.16(b)(3)(vii) (C)(1))   Yes___ No___

14. Wastes are not placed in an unwashed container that previously held an incompatible waste? (40 CFR 262.16(b)(3)(vii)(C)(2))   Yes___ No___

15. Tank(s) labeled as 'Hazardous Waste'? (40 CFR 262.16(b)(6)(ii)(A))   Yes___ No___

16. Tank(s) labeled with an indication of the nature of the hazard? (40 CFR 262.16(b) (6)(ii)(B))   Yes___ No___

17. Tanks are emptied within 180/270 days and documented using inventory logs, monitoring equipment or other records? (40 CFR 262.16(b)(6)(ii)(C))   Yes___ No___

18. Inventory logs or records demonstrating accumulation time are available for inspection? (40 CFR 262.16(b)(6)(ii)(D))   Yes___ No___

TABLE A.12
Record the following information for each tank:

Tank ID #____ Location of tank: _____

Tank design capacity: _____

Type of waste in tank: _____

Volume currently in the tank: _____

How was volume determined? _____

Person responsible for tank: _____

Age of tank when it first stored/treated/held a hazardous waste: _____

How is age of tank documented? _____

Tank ID #____ Location of tank: _____

Tank design capacity: _____

Type of waste in tank: _____

Volume currently in the tank: _____

How was volume determined? _____

Person responsible for tank: _____

Age of tank when it first stored/treated/held a hazardous waste: _____

How is age of tank documented? _____

TABLE A.13
Large Quantity Generator Hazardous Waste Checklist

Date: _____ Time: _____
Inspector's name: _____ Signature: _____

Records and Documentation

28. Facility continues to meet the Small Quantity Generator category >100 kg and
    <1,000 kg of hazardous waste and <1 kg of P-listed waste generated per
    month?   Yes___ No___
29. Waste is accumulated on-site for ≤180 days or ≤270 days if transporting
    greater than 200 miles if a non-permitted TSDF?   Yes___ No___
30. On-site accumulation of waste is <6,000 kg at all times?   Yes___ No___
31. Manifest system is used per 40 CFR 262.20(a)(1)?   Yes___ No___
32. Manifests are maintained for three years? (40 CFR 262.40(a))   Yes___ No___
33. Has obtained an EPA I.D. number? (40 CFR 262.20(a))   Yes___ No___
34. Manifests contain generator name, address and phone number? (262.20(a))   Yes___ No___
35. Transporter(s) name and EPA I.D. number recorded on manifest? (40 CFR
    262.20(a))   Yes___ No___
36. Designated facility name, address and EPA I.D. number included on manifest?
    (40 CFR 262.20(a))   Yes___ No___
37. Alternative facility designated on manifest (optional)? (40 CFR 262.20(c))   Yes___ No___
38. Are unique preprinted manifest tracking number and number of pages on
    manifest? (40 CFR 262.20(a))   Yes___ No___
39. DOT shipping name, hazard class, waste code and reportable quantity (RQ)
    (if required by 49 CFR 172) included on manifest? (40 CFR 262.20(a))   Yes___ No___
40. Number, type, quantity and unit wt./vol. of containers recorded on manifest?
    (40 CFR 262.20(a))   Yes___ No___
41. Proper certification including waste minimization recorded on manifest?
    (40 CFR 262.20(a))   Yes___ No___
42. Manifest are signed and dated by generator and transporter? (40 CFR 262.23(a)) Yes___ No___
43. Exception reports have been submitted if necessary? (40 CFR 262.42)   Yes___ No___
44. LDR notification/certification are sent with manifests on first shipment?
    (40 CFR 262.16(b)(7)→268.7(a)(2))   Yes___ No___
45. LDR notification/certification includes manifest number, correct EPA waste
    codes and treatment standards and waste analysis data? (40 CFR 262.16(b)
    (7)→268.7(a)(2))   Yes___ No___
46. LDR notification/certification/waste analysis data and other documents are
    maintained for three years? (40 CFR 262.16(b)(7)→268.7(a)(8))   Yes___ No___

TABLE A.14
Large Quantity Generator Hazardous Waste Checklist

Date: _____ Time: _____

Inspector's name: _____ Signature: _____

Waste Analysis/Waste Determination and Land Disposal Restrictions

19. Waste is not diluted impermissibly to meet LDR standards? (40 CFR262.16(b)(7)→268.3(a)
    & (b))   Yes___ No___
20. Are hazardous waste determinations made at the point of generation before any dilution,
    mixing or other alteration of the waste occurs and at any time in the course of its management
    where properties may have changed? (262.14(a)(2)→262.11(a))   Yes___ No___
21. Determinations made as to whether waste meets any of the listings in 40 CFR 261 Subpart D?
    (262.14(a)(2)→262.11(c))   Yes___ No___
22. Determinations made as to whether waste exhibits any of the characteristics identified in 40
    CFR 261 Subpart C? (262.14(a)(2)→262.11(d))   Yes___ No___
23. All applicable EPA hazardous waste numbers are identified? (40 CFR 262.11(g))   Yes___ No___
24. Records supporting hazardous waste determinations are maintained for at least three years
    from the date that the waste was last sent to on-site or off-site treatment, storage or disposal?
    (40 CFR 262.11(f))   Yes___ No___
25. Waste determinations including results of any tests, sampling, waste analyses or other
    determinations; tests, sampling and analytical methods; descriptions or processes, waste
    composition and waste properties; and records which explain the knowledge basis are
    documented? (40 CFR 262.11(f))   Yes___ No___
26. Are determinations made for waste that does not meet applicable treatment standards (ATS)?
    (40 CFR262.16(b)(7)→268.7(a)(2))   Yes___ No___
27. One-time written notice to TSDF has been submitted with initial shipment and a copy placed
    in file? (40 CFR 262.16(b)(7)→268.7(a)(2))   Yes___ No___
28. Is there waste covered by a National Capacity Variance(s)-268 Subpart C, Extension or
    Petition? (40 CFR 262.16(b)(7)→268.5 & 6)   Yes___ No___
    Describe the variance, extension or petition that applies
29. For waste shipped off-site for disposal, has a notice been provided to the land disposal facility
    with the initial shipment or a revised notice if changes occur, stating that the waste is exempt
    from the LDRs? (40 CFR 262.16(b)(7)→268.7(a)(4))   Yes___ No___
30. For waste shipped off-site for disposal, has a notice been provided with the initial shipment or
    new notification if changes occur? (40 CFR 262.16(b)(7)→268.7(a)(2))   Yes___ No___
31. For waste shipped off-site for disposal, are the following included on the LDR notice: EPA
    hazardous waste number(s), manifest number(s), waste analysis data, if available, and waste
    constituents, wastewater or non-wastewater classification and subcategory, if applicable? (40
    CFR 262.16(b)(7)→268.7(a)(2) and (a)(4))   Yes___ No___
32. If waste is determined to be excluded from the definition of hazardous or solid waste, is a
    one-time notice describing generation, subsequent exclusion or exemption and disposition of
    the waste retained in the facility's on-site files? (40 CFR 262.16(b)(7)→268.7(a)(7))   Yes___ No___
33. If generator determines waste or soil contaminated with waste does meet the ATS or does not
    exceed prohibition levels and requires no further treatment, has a one-time written notice been
    submitted to the TSDF with initial shipment and a copy placed in file? (40 CFR 262.16(b)
    (7)→268.7(a)(3)(i))   Yes___ No___
34. Have underlying hazardous constituents as defined in 40 CFR 268.2(i) been determined for
    waste that is not D001 non-wastewater? (40 CFR 262.16(b)(7)→268.9(a))   Yes___ No___
35. For waste that is land disposed, has it been determined if waste meets the treatment standards
    specified in 268 Subpart D? (40 CFR 262.16(b)(7)→268.9(c))   Yes___ No___
36. For claims that a characteristic waste is no longer hazardous, has a one-time notification and
    certification been sent to EPA or authorized state, a copy placed in the file and updated if there
    are changes in process, operation or receiving facility? (40 CFR 262.16(b)(7)→268.9(d))   Yes___ No___

TABLE A.15

Hazardous Waste Satellite Accumulation Area Checklist

Date: _____ Time: _____

Inspector's name: _____ Signature: _____

13. Is SAA at or near the point of generation? --- ☐ Yes ☐ No

14. Is SAA under control of the responsible operator? -- ☐ Yes ☐ No

15. Is quantity of waste less than 55 gals (or 1 kg/1 quart of P-listed waste)? ---------------- ☐ Yes ☐ No

16. Waste exceeding above quantities in SAA transferred within three days? ----------------- ☐ Yes ☐ No

17. Containers labeled with words 'Hazardous Waste' (262.15(a)(5)(i))? ---------------------- ☐ Yes ☐ No

18. Containers labeled with nature of the hazard (262.15(a)(5)(ii))? ---------------------------- ☐ Yes ☐ No

19. Are containers kept closed when not adding waste?--- ☐ Yes ☐ No

20. Are containers in good condition? --- ☐ Yes ☐ No

21. Are containers compatible with waste being stored?-- ☐ Yes ☐ No

22. Incompatible waste not being placed into the same container?------------------------------- ☐ Yes ☐ No

23. Wastes are not placed in an unwashed container that previously held an incompatible
    waste?-- ☐ Yes ☐ No

24. Containers of incompatible waste are kept separated? -- ☐ Yes ☐ No

Describe any issues identified during the inspection and corrective actions taken:

SAA ID:_ SAA Location: _____

Person responsible for SAA: _____

Description of problems: _____

Actions taken: _____

SAA ID:_ SAA location: _____

Person responsible for SAA: _____

Description of problems:_____

Actions taken: _____

SAA ID:_ SAA location: _____

Person responsible for SAA: _____

Description of problems:_____

Actions taken: _____

TABLE A.16
Large Quantity Generator Hazardous Waste Checklist

Date: _____ Time: _____

Inspector's name: _____ Signature:_____

<u>Container Accumulation Area</u>

  9. Containers being used are in good condition? (40 CFR 262.17(a)(1)(ii))   Yes___ No___

10. Containers being used are compatible with waste? (40 CFR 262.17(a)(1)(iii))   Yes___ No___

11. Containers closed are kept closed? (40 CFR 262.17(a)(1)(iv)(A))   Yes___ No___

12. Containers are not opened, handled or managed in a manner to cause them to leak? (40 CFR 262.17(a)(1)(iv)(B))   Yes___ No___

13. CAA is inspected weekly? (40 CFR 262.17(a)(1)(v))   Yes___ No___

14. Containers of ignitable/reactive waste are kept more than 50 feet from property line? (40 CFR 262.17(a)(1)(vi)(A))   Yes___ No___

15. Measures are taken to prevent accidental ignition or reaction of ignitable/reactive waste, including separation from ignition sources, smoking/open flame restrictions and 'No Smoking' signs? (40 CFR 262.17(a)(1)(vi)(B))   Yes___ No___

16. Incompatible wastes are not placed in the same container? (40 CFR 262.17(a)(1)(vii)(A))   Yes___ No___

17. Wastes are not placed in an unwashed container that previously held an incompatible waste? (40 CFR262.17(a)(1)(vii)(B))   Yes___ No___

18. Containers of incompatible waste are separated or otherwise protected from each other? (40 CFR 262.17(a)(1)(vii)(C))   Yes___ No___

19. Containers are labeled as 'Hazardous Waste'? (40 CFR 262.17(a)(5)(i)(A))   Yes___ No___

20. Containers are labeled with an indication of the nature of the hazard? (40 CFR 262.17(a)(5)(i)(B))   Yes___ No___

21. Containers are marked with accumulation start dates? (40 CFR 262.17(a)(5)(i)(C))   Yes___ No___

PRE-TRANSPORT REQUIREMENTS

22. Containers are packed, labeled and marked per DOT requirements? (40 CFR 262.30, 262.31,262.32, respectively)   Yes___ No___

23. Placards are provided for use by transporters when applicable? (40 CFR 262.33)   Yes___ No___

24. Containers are marked with all applicable EPA hazardous waste numbers? (40 CFR 262.11(g))   Yes___ No___

TABLE A.17
Large Quantity Generator Hazardous Waste Checklist

Date: _____ Time: _____
Inspector's name: _____ Signature: _____
Preparedness and Prevention

25. An emergency coordinator has been designated and that person is on-site or
on-call? (40 CFR 262.17(a)(6)-> 262.264)   Yes___ No___

26. Emergency coordinator is thoroughly familiar with all aspects of the generator's
operations? (40 CFR 262.17(a)(6)→262.264)   Yes___ No___

27. Arrangements have been made with local emergency agencies? (40 CFR 262.17(a)
(6)→262.256(a))   Yes___ No___

28. Emergency agencies have been familiarized with the layout of the facility,
properties of hazardous waste handled at the facility, normal work locations,
entrances and roads inside the facility, possible evacuation routes and typical
injuries or illnesses which could result from an incident? (40 CFR 262.17(a)
(6)→262.256(a)(2))   Yes___ No___

29. Primary response entity has been designated if more than one police or fire
department might respond? (40 CFR 262.17(a)(6)→262.256(a)(3))   Yes___ No___

30. Records documenting arrangements with response agencies are maintained or, in
cases where no arrangements exist, there is proof that attempts to make such
arrangements were made? (40 CFR 262.17(a)(6)→262.256(b))   Yes___ No___

31. If 24-hour response capabilities exist in lieu of arrangements, has a waiver of
arrangements with emergency response agencies been obtained? (40 CFR
262.17(a)(6)→262.256(c))   Yes___ No___

TABLE A.18
Large Quantity Generator Hazardous Waste Checklist

Date: _____ Time: _____
Inspector's name: _____ Signature: _____

Response Equipment

32. Facility operates to minimize possibility of a fire, explosion or release? (40 CFR
    262.17(a)(6)→262.251)   Yes___ No___
33. An internal communications or alarm system is provided? (40 CFR 262.17(a)
    (6)→262.252(a))   Yes___ No___
34. A device is available for and capable of summoning emergency assistance? (40
    CFR 262.17(a)(6)→262.252(b))   Yes___ No___
35. Adequate supply and proper spill control, decontamination and safety equipment
    (fire blankets, respirators, absorbent, etc.) are provided? (40 CFR 262.17(a)
    (6)→262.252(c))   Yes___ No___
36. Adequate water supply for fire control equipment is provided? (40 CFR 262.17(a)
    (6)→262.252(d))   Yes___ No___
37. Communication and emergency equipment are tested? (40 CFR 262.17(a)
    (6)→262.253)   Yes___ No___
38. Communication equipment is immediately accessible when waste is being
    handled? (40 CFR 262.17(a)(6)→262.254(a))   Yes___ No___
39. Communication equipment is immediately accessible when only one person
    present on-site? (40 CFR 262.17(a)(6)→262.254(b))   Yes___ No___
40. Adequate aisle space is provided for type of waste management and emergency
    equipment used? (40 CFR 262.17(a)(6)→262.255)   Yes___ No___

TABLE A.19

Large Quantity Generator Hazardous Waste Checklist

Date:_____ Time:_____

Inspector's name: _____ Signature: _____

Contingency Planning

41. Facility has a contingency plan? (40 CFR 262.17(a)(6)→262.260(a))   Yes___ No___
42. Contingency plan is reviewed and amended when (a) regulations are revised; (b) the plan fails in an emergency; (c) the generator's operations materially change to increase risk of fire, explosions or releases; (d) the ECs change; or (e) the emergency equipment changes? 40 CFR 262.17(a)(6)→262.263(a) through 263(e))   Yes___ No___
43. Contingency plan has been submitted to emergency response agencies? (40 CFR 262.17(a)(6)→262.262(a))   Yes___ No___
44. Contingency plan includes description of actions needed to respond to fires, explosions or releases of hazardous wastes? (40 CFR 262.17(a)(6)→262.261(a))   Yes___ No___
45. Contingency plan includes description of arrangements with local emergency agencies, as appropriate? (40 CFR262.17(a)(6)→262.261(c))   Yes___ No___
46. Contingency plan identifies EC(s) by name (or by position title for 24-hour facilities) and lists emergency telephone number(s)? (40 CFR 262.17(a)(6)→262.261(d))   Yes___ No___
47. List of ECs is kept up to date? (40 CFR -262.17(a)(6)→262.261(d))   Yes___ No___
48. If more than one EC, one is designated as primary and lists the others in the order they will assume responsibility as alternatives? (40 CFR 262.17(a)(6)→262.61(d))   Yes___ No___
49. Contingency plan includes list and describes emergency equipment, its location and its capabilities? (40 CFR 262.17(a)(6)→262.261(e))   Yes___ No___
50. Contingency plan includes complete evacuation plan, including route, signal and alternative route (if required)? (40 CFR 262.17(a)(6)→262.261(f))   Yes___ No___
51. A quick reference guide has been prepared and submitted to emergency response agencies (new facilities beginning operations after May 30, 2017, or facilities revising the contingency plan after that date)? (40 CFR 262.17(a)(6)→262.262(b))   Yes___ No___
52. The quick reference guide includes (1) list and description of hazardous waste described in layman's terms; (2) estimated maximum amounts of each hazardous waste; (3) identification of any hazardous waste that would result in unique or special medical issues; (4) a map showing where hazardous waste is generated or managed and access routes to these locations; (5) a map of the facility in relation to surrounding businesses, schools and residential areas to allow access and evacuation planning; (6) locations of water supplies; (7) identification of on-site notification systems or alarms; and (8) name and 24/7 telephone number for the EC? (40 CFR 262.17(a)(6)→262.262(b)(1) through (b)(8))   Yes___ No___
53. The quick reference guide is updated and resubmitted to emergency response agencies when needed? (40 CFR 262.17(a)(6)→262.262(c))   Yes___ No___

TABLE A.20
Large Quantity Generator Hazardous Waste Checklist

Date:_____ Time: _____
Inspector's name: _____ Signature:_____

Personnel Training

54. Hazardous waste personnel are trained to perform their duties in a way that ensures compliance using a program of classroom instruction, online training (e.g., computer-based or electronic) or on-the-job training? (40 CFR 262.17(a)(7)(i)(A))   Yes___ No___

55. Instructor(s) are trained in hazardous waste management procedures? (40 CFR 262.17(a)(7)(i)(B))   Yes___ No___

56. Hazardous waste training includes response to emergencies, implementation of contingency plan, use of alarms, waste feed cutoffs and other emergency equipment, as required? (40 CFR 262.17(a)(7)(i)(C))   Yes___ No___

57. New employees are initially trained within six months of employment or assignment to the generator or position? (40 CFR 262.17(a)(7)(ii))   Yes___ No___

58. Hazardous waste training is refreshed annually? (40 CFR 262.17(a)(7)(iii))   Yes___ No___

59. Specific job titles and names of persons filling positions are tracked? (40 CFR 262.17(a)(7)(iv)(A))   Yes___ No___

60. Skills, education or qualification and duties associated with each job title are described in writing? (40 CFR 262.17(a)(7)(iv)(B))   Yes___ No___

61. Written description of type and amount of introductory and continuing training has been prepared? (40 CFR 262.17(a)(7)(iv)(C))   Yes___ No___

62. Documentation confirming training has been completed is maintained? (40 CFR 262.17(a)(7)(iv)(D))   Yes___ No___

63. Training records of current employees and former employees are maintained for three years? (40 CFR 262.17(a)(7)(v))   Yes___ No___

TABLE A.21

Large Quantity Generator Hazardous Waste Checklist

Date: _____ Time: _____
Inspector's name: _____ Signature: _____

Tank Accumulation Area

Existing tanks (installation prior to July 14, 1986)

64. All existing tanks have been assessed by an Independent PE? 40 CFR 262.17(a)(2)→265.191(a))

65. Existing systems that store material which became hazardous waste subsequent to
    July 14, 1986, have been assessed? (40 CFR 262.17(a)(2)→265.191(c))

66. Assessment includes the following: design standards, characteristics of waste,
    existing corrosion protection, age, leak test for nonenterable tanks and ancillary
    equipment? (40 CFR 262.17(a)(2)→265.191(b))   Yes___ No___

New tanks (installation after July 14, 1986)

67. All existing tanks have been assessed by an independent PE? (40 CFR 262.17(a)
    (2)→265.192(a))   Yes___ No___

68. Assessment includes the following: design standards, characteristics of waste,
    corrosion protection (completed by corrosion expert), tightness prior to use? (40 CFR
    262.17(a)(2)→265.192(a)(1-5))   Yes___ No___

69. Independent PE has performed installation inspection? (40 CFR 262.17(a)
    (2)→265.192(b))   Yes___ No___

70. Certification statements of design and inspection are maintained? (40 CFR 262.17(a)
    (2)→265.192(g))   Yes___ No___

All hazardous waste tanks

71. Tanks are labeled 'Hazardous Waste'? (40 CFR 262.17(a)(5)(ii)(A))   Yes___ No___

72. Tanks are labeled with an indication of the nature of the hazard? (40 CFR 262.17(a)
    (5)(ii)(B))   Yes___ No___

73. Inventory logs, monitoring equipment or other records are maintained to demonstrate
    that hazardous waste has been emptied within 90 days (batch process), *or* other
    means are provided to demonstrate that estimated volumes of hazardous
    waste entering the tank daily exit within 90 days (continuous process)?
    (40 CFR2 62.17(a)(5)(ii)(C))   Yes___ No___

74. Inventory logs or records to demonstrate 90-day compliance on-site are maintained
    and readily available? (40 CFR 262.17(a)(5)(ii)(D))   Yes___ No___

75. Secondary containment (sec. cont.) is installed for all tanks in the following
    categories: new; installed after July 14, 1986; over 15 years old or unknown age in
    facility over 15 years old; repaired, replaced or reinstalled after July 14, 1986? (40
    CFR 262.17(a)(2)→265.193(a))   Yes___ No___

76. Secondary containment material is constructed of impervious and compatible
    material? (40 CFR 262.17(a)(2)→265.193(c)(1))   Yes___ No___

77. Secondary containment is capable of preventing failure due to settlement,
    compression or uplift? (40 CFR 262.17(a)(2)→265.193(c)(2))   Yes___ No___

78. Ancillary equipment is placed in secondary containment, except aboveground piping,
    welded flanges, joints, connections, sealless or magnetic pumps, pressurized piping
    with automatic shutoff devices, if inspected daily? (40 CFR 262.17(a)
    (2)→265.193(f))   Yes___ No___

79. Leak detection system is installed for secondary containment capable of detecting
    leaks within a 24-hour period? (40 CFR 262.17(a)(2)→265.193(c)(3))   Yes___ No___

80. Spilled or leaked waste and precipitation are removed from secondary containment
    within 24 hr. or as soon as possible? (40 CFR 262.17(a)(2)→265.193(c)(4))   Yes___ No___

81. Secondary containment is capable of containing 100% of largest tank? (40 CFR
    262.17(a)(2)→265.193(e)(1)(i))   Yes___ No___

82. Spill and overflow prevention controls are installed and operated (check valves,
    dry disconnects, level-sensing devices, high-level alarms, automatic feed
    cutoffs, maintenance of sufficient freeboard, etc.)? (40 CFR 262.17(a)
    (2)→265.194(b))   Yes___ No___

TABLE A.21 (CONTINUED)
Large Quantity Generator Hazardous Waste Checklist

83. Tanks are compatible with waste or treatment method? (40 CFR 262.17(a)
    (2)→265.194(a))   Yes___ No___
84. Incompatible wastes are not placed in the same tank? (40 CFR 262.17(a)
    (2)→265.199(a))   Yes___ No___
85. Ignitable/reactive wastes are treated and accumulated per NFPA's buffer zone
    requirements? (40 CFR 262.17(a)(2)→265.198(b))   Yes___ No___
86. Ignitable/reactive wastes are treated and accumulated so as to prevent ignition?
    (40 CFR 262.17(a)(2)→265.198(a))   Yes___ No___
87. The following are inspected daily: spill/overflow equipment, aboveground portions
    of tank system, sec. cont. and data from monitoring equipment? (45 CFR 262.17(a)
    (2)→265.195(a))   Yes___ No___
88. Cathodic protection systems are inspected annually and impress current systems
    every two months? (40 CFR 262.17(a)(2)→265.195(b))   Yes___ No___

Record the following information for each tank:
Tank ID #____ Location of tank:
Tank design capacity:
Type of waste in tank:
Volume currently in the tank:
How was volume determined?
Person responsible for tank:
Age of tank when it first stored/treated/held a hazardous waste:
How is age of tank documented:
Tank ID #____ Location of tank:
Tank design capacity:
Type of waste in tank:
Volume currently in the tank:
How was volume determined?
Person responsible for tank:
Age of tank when it first stored/treated/held a hazardous waste:
How is age of tank documented:

TABLE A.22

Used Oil Checklist (for VSQG, SQG and LQG)

Note: For very small quantity generators, mixtures of used oil and very small quantity generator hazardous waste regulated under §262.14 of this chapter are subject to regulation as used oil (40 CFR 279.10 (b)(3)).

Date: _____ Time: _____

Inspector's name: _____ Signature: _____

1. Mixtures of hazardous waste and used oil are managed according to 279.10(b)? Yes___ No___
2. Rebuts the presumption that listed hazardous waste has been mixed with used oil for containing more than 1,000 ppm total halogens? Yes___ No___
3. Used oil is stored only in tanks, containers or units subject to regulation under 40 CFR Parts 264 or 265 per 40 CFR 279.22(a)? Yes___ No___
4. Used oil is stored in containers and ASTs that are (1) in good condition and (2) have no visible leaks per 279.22(b)(1) and (b)(2)? Yes___ No___
5. Containers and ASTs are labeled 'Used Oil' per 40 CFR 279.22(c)(1)? Yes___ No___
6. Labels or marks fill pipes used for underground tanks as 'Used Oil' per 279.22(c) (2)? Yes___ No___
7. When a release is detected, (1) the release is stopped, (2) the release is contained, (3) used oil and other materials are cleaned up and managed and (4) the containers or tanks are repaired or replaced prior to returning them to service, if necessary, per 40 CFR 279.22(d)(1) through (d)(4)? Yes___ No___
8. Any used oil burned on-site is limited to facility generated or household DIY used oil? Yes___ No___
9. Any used oil burned on-site is burned in a <0.5 M BTU/hr. space heater that is vented to ambient air? Yes___ No___
10. Per 40 CFR 279.24, used oil is transported only by a transporter who has obtained an EPA I.D. number? Yes___ No___
11. If tolling agreement is in place, does the contract include the following: (1) type of used oil and frequency of shipments, (2) requirement that the vehicle transporting the used oil to and from generator is owned by the processor/re-refiner and (3) requirement that the reclaimed oil will be returned to generator per 279.24(c)(1) through (c)(3)? Yes___ No___
12. If transporting facility–generated used oil is used in our own vehicles, are quantities less than 55 gallons at a time and transported to a recognized used oil collection center per 40 CFR-279.24(a)? Yes___ No___
13. If transporting facility–generated used oil is used in our own vehicles, are quantities less than 55 gallons at a time and transported to a company-owned aggregation point 40 CFR 279.24(b)? Yes___ No___
14. For any company-owned used oil aggregation point, are all registrations and licenses required by state and/or local government in place and current? Yes___ No___

TABLE A.23
Universal Waste Checklists (For VSQG, SQG and LQG)

Note: Very small quantity generators have the option to manage their universal waste items as hazardous waste or under the universal waste requirements in 40 CFR 273.

Date: _____ Time: _____

Inspector's name: _____ Signature: _____

1. Does not dispose of universal waste on-site per 40 CFR 273.11(a)? Yes___ No___
2. Does not dilute or treat universal waste, except for responding to releases per 273.17 or by managing specific wastes per 273.13 (waste management), per 40 CFR 273.11(b)? Yes___ No___
3. Does not accumulate universal waste for longer than one year per 40 CFR 273.15(a)? Yes___ No___
4. Demonstrates the length of time that the universal waste has been accumulated per 40 CFR 273.15(c)? Yes___ No___
5. Trains employees responsible for management of universal waste in proper handling and emergency procedures per 40 CFR 273.16? Yes___ No___
6. Immediately contains all releases of universal wastes and other residues from universal wastes per 40 CFR 273.17(a)? Yes___ No___
7. Makes a hazardous waste determination on any materials resulting from a release or from any materials (such as electrolytes) generated from management of universal waste? Yes___ No___
8. Is universal waste transported to a universal waste handler or destination facility that is subject to the requirements of 40 CFR 273? Yes___ No___

Universal Waste Lamps
1. Universal waste lamps are kept in containers or packages that are closed, structurally sound, compatible and lack evidence of leakage, spillage or damage that could cause leakage per 40 CFR 273.13(d)(1)? Yes___ No___
2. Universal waste lamps that show evidence of breakage or damage are immediately contained per 40 CFR 273.13(d)(2)? Yes___ No___
3. Containers of universal waste lamps are labeled: 'Universal Waste-Lamp(s)' or 'Waste Lamp(s)' or 'Used Lamp(s)' per 40 CFR 273.14(e)? Yes___ No___

Universal Waste Batteries
1. Universal waste batteries that show any evidence of leakage or other damage are immediately contained per 40 CFR 273.13(a)(1)? Yes___ No___
2. Individual batteries or their containers are labeled with 'Universal Waste-Battery(ies)' or 'Waste Battery(ies)' or 'Used Battery(ies)' per 40 CFR 273.14(a)? Yes___ No___

Universal Waste Mercury-Containing Equipment (MCE)
1. Universal waste MCE that show any evidence of leakage or other damage is immediately contained per 40 CFR 273.13(c)(1)? Yes___ No___
2. Mercury ampules from MCE is removed only with all health and safety requirements in place per 40 CFR 273.13(c)(2)(i) through (2)(vi)? Yes___ No___
3. Mercury ampule from MCE is stored in closed containers, packed to prevent breakage per 40 CFR 273.13(c)(2)(vii) through (2)(viii)? Yes___ No___
4. Individual MCE or their containers are labeled 'Universal Waste-Mercury Containing Equipment' or 'Waste Mercury-Containing Equipment' or 'Used Mercury- Containing Equipment' per 40 CFR 273.14(d)(1)? Yes___ No___
5. Individual MCE thermostats or their containers are labeled 'Universal Waste-Mercury Thermostat(s)' or 'Waste Mercury Thermostat(s)' or 'Used Mercury Thermostat(s)' per 40 CFR 273.14(d)(2)? Yes___ No___

Universal Waste Pesticides
1. Universal waste pesticides are kept in containers that are closed, structurally sound, compatible and lack evidence of leakage, spillage or damage that could cause leakage per 40 CFR 273.13(b)(1)? Yes___ No___
2. Universal waste pesticides that are managed in a tank, does tank meet requirements of 40 CFR 265 Subpart J per 40 CFR 273.13(b)(1)? Yes___ No___
3. Universal waste pesticide in noncompliant containers are overpacked in a container compliant with 40 CFR 273.13(b)(1)? Yes___ No___
4. Universal waste pesticides are kept in a transport vehicle/vessel that is closed, structurally sound, compatible and lacks evidence of leakage, spillage or damage which could cause leakage per 40 CFR 273.13(b)(4)? Yes___ No___
5. Recall universal waste pesticides are labeled (1) with the original product label or appropriate DOT label as identified in 49 CFR 172 and (2) 'Universal Waste-Pesticide(s)' or 'Waste-Pesticide(s)' per 40 CFR 273.14(b)(1) and (b)(2)? Yes___ No___
6. Unused pesticide products are labeled with at least one of the following: (i) the label that was on the product when purchased, if still legible; (ii) the appropriate label required under DOT regulation; or (iii) another label prescribed or designated by the state waste pesticide collection program per 40 CFR 273.14(c)(1)(i) through (1)(iii)? Yes___ No___
7. Unused pesticide products are labeled with 'Universal Waste-Pesticide(s)' or 'Waste-Pesticide(s)' per 40 CFR 273.14(c)(1)(2)? Yes___ No___

TABLE A.24

Land Disposal Restrictions (LDRs) Checklist

Date: _____ Time: _____

Inspector's name: _____ Signature: _____

Note: This checklist is for SQGs and LQGs that manage the following hazardous waste subject to LDRs: contaminated soil, or contaminated soil which is treated, or a lab pack waste, or hazardous waste debris, or managed at a treatment or disposal facility or where the generator's determination is based solely on knowledge.

1. Waste or soil contaminated with waste has been determined to meet the ATS or does not exceed prohibition levels and requires no further treatment? (40 CFR 268.7(a)(3))   Yes___ No___

2. One-time written notice has been submitted to treatment or storage facility with initial shipment and a copy placed in file? (40 CFR 268.7(a)(3)(i))   Yes___ No___

3. For soil contaminated with waste, a special certification statement was included with the notice? (40 CFR 268.7(a)(2))   Yes___ No___

4. Waste analysis plan has been developed for treatment in tanks/containers to meet LDR treatment standards found in 268.40? (40 CFR 268.7(a)(5))   Yes___ No___

5. If shipping off-site, a notice and certification statement has been provided with initial shipment, or renotification if the waste changes? (40 CFR 268.7(a)(3)-268.7(a)(5)(iii))   Yes___ No___

6. For waste or contaminated soil determined to be restricted based solely on knowledge of the waste, all supporting data is maintained on-site in generator's files? (40 CFR 268.7(a)(6))   Yes___ No___

7. For waste or contaminated soil determined to be restricted based on testing the waste or an extract, all supporting data is maintained on-site in generator's files? (40 CFR 268.7(a)(6))   Yes___ No___

8. Uses the alternative treatment standards to manage lab pack containing hazardous waste(s)? (40 CFR 268.7(a)(9))   Yes___ No___

9. A notice and certification is provided with the initial shipment for lab pack waste with ATS? (40 CFR 268.7(a)(9)(i))   Yes___ No___

10. New notice and certification is provided when changes in waste or receiving facility occur? (40 CFR 268.7(a)(9)(ii))   Yes___ No___

11. For waste determined to be excluded from the definition of hazardous or solid waste, or exempt from Subtitle C regulations under 261.2 through 261.6 subsequent to the point of generation, retains a one-time notice describing generation, subsequent exclusion or exemption and disposition of the waste in the facility's on-site files? (40 CFR 268.7(a)(7))   Yes___ No___

12. For hazardous debris determined to be excluded from the definition of hazardous waste under 261.3(e), submits one-time notification to EPA or authorized state? (40 CFR 268.7(d)(1))   Yes___ No___

13. For hazardous debris determined to be excluded from the definition of hazardous waste under 261.3(e), the notification has been updated if debris is shipped to another location, if different debris is treated or if a different technology is used for treatment? (40 CFR 268.7(d)(2))   Yes___ No___

14. For contaminated soil subject to alternative LDR treatment standards in 268.49(a), has determined or received a determination from EPA or an authorized state that such contaminated soil no longer exhibits a characteristic of hazardous waste? (40 CFR 268.7(e))   Yes___ No___

15. One-time-only documentation of these determinations has been prepared, including all supporting information? (40 CFR 268.7(e)(1))   Yes___ No___

16. The one-time documentation of the determinations, including all supporting information and other records, is maintained for a minimum of three years? (40 CFR 268.7(e)(2))   Yes___ No___

TABLE A.25
Process Vents (Subpart AA) Checklist

Date: _____ Time: _____

Inspector's name: _____ Signature: _____

Note: This checklist is for TSDFs and LQGs that manage hazardous wastes containing greater than 10 ppmw of organics in a process vent used in distillation, fractionation, solvent extraction, thin-film evaporation and air or steam stripping.

1. Does the facility have any hazardous waste management unit(s) using the following processes: distillation, fractionation, thin-film evaporation, solvent extraction, air stripping and steam stripping? If *no*, proceed to the Subpart BB checklist.   Yes___ No___
2. Are any of these processes exempt under the closed-loop recycle exemption?   Yes___ No___
1. Explain any exemptions
3. Does the hazardous waste contain greater than 10 ppmw organics?   Yes___ No___
4. For those process vents with greater than 10 ppmw organics, describe the waste(s), unit(s) and processes.
5. For process vents with less than 10 ppmv organics, describe the information/ documentation used to make the determination.
6. Does the total hourly emission rate of process vents with greater than 10 ppmv organics exceed 3 lb/hr.? Does the facility-wide yearly emission rate exceed 3.1 tons/year?   Yes___ No___
7. If either answer to #6 above is No, are the calculations to support this determination documented and available?   Yes___ No___
8. If either answer to #6 above is Yes, are control devices installed to reduce the emissions? (*All TSDs must have control devices in place.*)   Yes___ No___
9. Are the calculations/analysis current?   Yes___ No___
10. Are facility operating hours (e.g., 8 or 24 hr./day) correct?   Yes___ No___
11. Have worst-case scenarios been considered?   Yes___ No___
12. Are control devices inspected and/or monitored at least once each operating day to ensure proper operation? (40 CFR 264/5.1035(c))   Yes___ No___
13. Is there any indication of a problem with operation of control devices?   Yes___ No___
14. Were corrective measures implemented immediately for any problems that have arisen?   Yes___ No___
15. Are design data and monitoring data for control devices documented and available?   Yes___ No___

TABLE A.26
Equipment Leaks (Subpart BB) Checklist

Date: _____ Time:_____

Inspector's name:_____ Signature:_____

Note: This checklist is for TSDFs and LQGs that have equipment (any valve, pump, compressor, pressure relief device, sampling connection system, flange, open-ended valve or line) that contacts hazardous wastes containing greater than 10% organics.

For pumps and valves in light liquid or gas/vapor service: For a hazardous waste to be in light liquid service, the vapor pressure of one or more organic constituents in the material must exceed 0.3 kilopascals at 20°C and total concentration of pure organic constituents exerting vapor pressure exceeding 0.3 kilopascals at 20°C must equal or exceed 20% by weight.

1. Is each pump in light liquid service monitored monthly to detect leaks? (40 CFR 264/5.1052(a)(l))   Yes___ No___

2. Is each pump in light liquid service checked by visual inspection each calendar week for indications of liquids dripping from the pump seal? (40 CFR 264/5.1052(a)(2))   Yes___ No___

3. Is each valve in light liquid service or gas/vapor service monitored monthly for leaks?(40 CFR 264/5.1057(a))   Yes___ No___

Equipment in heavy liquid service

1. Are pumps, valves in heavy liquid service, pressure relief devices in light liquid or heavy liquid service and flanges and other connectors in light liquid or heavy liquid service monitored for leaks by visual, olfactory or any other detection method?   Yes___ No___

TABLE A.27
Tanks and Containers (Subpart CC) Checklist

Date: _____ Time: _____

Inspector's name: _____Signature:_____

Note: This checklist is for TSDFs and LQGs that manage in tanks and containers hazardous waste containing volatile organic concentrations averaging 500 ppmw or more on an annual basis.

4. Are any units at the facility subject to the CC rule?   Yes___ No___
5. If the answer is no, what is the reason (mark applicable exemption below).
   a. 40 CFR 264/265.1080(b) exemptions:
      i. (1) Unit did not receive HW after 12/6/96 ___
      ii. Using containers of less than 26 gallons capacity___
      iii. Unit undergoing closure ___
      iv. Units used in an on-site RCRA or CERCLA cleanup___
      v. Mixed radioactive and hazardous waste___
      vi. Units with CAA, NESHAPS or NSPS controls___
      vii. Tanks with process vents (subject to Subpart AA)___
   b. 40 CFR 265.1083(c) exemptions:
      i. Waste stream less than 500 ppmw average VOCs.___ If so, was waste determination conducted per 265.1084?   Yes ___ No ___
      ii. All waste placed in unit meets 268.40 (LDR) limits___
      iii. Tank is used for bulk feed to incinerator and requirements of 265.1083(5)(i)-(iii) are met___
   c. 40 CFR 265.1 General exclusions/exemptions: ____
      i. Hazardous waste recycling unit exemption ___
      ii. Satellite accumulation area___
      iii. Totally enclosed treatment facility exemption ___
      iv. Elementary neutralization unit___
      v. Wastewater treatment in tanks exemption___
      vi. Emergency or spill management exemption___
      vii. Biological treatment with 95% efficiency. Except if exemption is based on waste stream less than 500 ppmw, STOP because Subpart CC does not apply._____
6. Is average volatile organic concentration in each waste management unit exceeding 500 ppmw determined on an average annual basis at the point of waste origination?   Yes___ No___
7. If claiming that volatile organic concentration in its waste is below 500 ppmw, is waste determination documentation in the operating record?   Yes___ No___
8. For units subject to Subpart CC, is there a list of each unit and the concentration in the facility operating record? (40 CFR 264.1089/265.1090)   Yes___ No___
9. For any units determined to not be subject to Subpart CC, is there a determination for each unit in the facility operating record? (40 CFR 264.1089/265.1090)   Yes___ No___

For each hazardous waste unit that has been determined to contain less than 500 ppmw:
10. How was waste determination conducted? Process knowledge___ or sampling___
11. If process knowledge was used, is any documentation on file?   Yes___ No___
12. Is the documentation adequate?   Yes___ No___
13. If sampling was used, is there a written sampling plan?   Yes___ No___
14. If sampling was used, was the sampling performed according to an EPA-approved method? List EPA method used:
15. Has the waste stream changed since the initial waste determination in a manner altering the character of the waste or causing exceedance of threshold levels for applicability of Subpart CC?   Yes___ No___
16. If waste stream has changed, was a new waste determination conducted and documented?   Yes___ No___

Containers (40 CFR 264.1086/265.1087 Light Liquid Service): For a hazardous waste to be in light liquid service, the vapor pressure of one or more of the organic constituents in the material must exceed 0.3 kilopascals at 20°C and total concentration of pure organic constituents must exert a vapor pressure exceeding 0.3 kilopascals at 20°C and must equal or exceed 20% by weight.

(Continued)

TABLE A.27 (CONTINUED)
Tanks and Containers (Subpart CC) Checklist

1. LEVEL ONE: There should be no waste stabilization. Containers must be >0.1m³ (26.4 gal) and ≤122 gallons. If the organic waste is not in light liquid service, its volume can exceed 122 gallons.
   a. OPTION 1: The container meets DOT specifications.
   b. OPTION 2: Use a cover and closure device on the container and ensure absence of visible gaps in the interior of the container or holes in the covers.
   c. OPTION 3: Use vapor suppressing barrier on or above the hazardous waste in the container.
2. LEVEL TWO: There should be no waste stabilization. Containers are larger than 0.46 m³ (122 gal) and are in light liquid service.
   a. OPTION 1: The container meets DOT specifications.
   b. OPTION 2: Operates with no detectible emissions from the container under Method 21.
   c. OPTION 3: Demonstrated to be vapor tight within the last 12 months via application of Method 27.
3. LEVEL THREE: Container must be used for waste stabilization. Vent vapors from containers and remove or destroy them in a control device. Put container in a 'Procedure T Enclosure,' vent vapors and destroy them in a control device.
4. Is the level of control imposed at the facility in compliance? (**Note:** Facility is in compliance if not conducting waste stabilization and waste is stored in DOT-approved, 55-gallon drums.)

Tanks

(For tank storage, a facility may manage its waste at two levels: Tank Level 1 requires a fixed roof to maintain a maximum organic vapor pressure complying with Subpart CC. Tank Level 2 design options number five: (1) an internal floating roof, (2) an external floating roof, (3) a tank with a fixed roof vented through a closed vent system to a control device, (4) a pressure tank and (5) a tank located inside an enclosure that is vented through a closed vent system to an enclosed combustion device.)

1. Is hazardous waste with average volatile organics concentration exceeding 500 ppmw placed in a tank with either Level 1 or Level 2 controls? (40 CFR 264.1084(b)(1)/265.1085(b)(1))   Yes___ No___
2. Were the tanks inspected for leaks before waste was placed into the tanks?   Yes___ No___
   Date of tank inspection_____
3. For tanks storing hazardous waste, was an annual inspection of the tanks conducted?
   Date of annual tank inspection

Note: The fixed roof and its closure devices shall be visually inspected to check for defects that could result in air pollutant emissions. Defects include, but are not limited to, visible cracks, holes or gaps in the roof sections or between the roof and the tank walls; broken, cracked or otherwise damaged seals or gaskets on closure devices; and broken or missing hatches, access covers, caps or other closure devices. An initial inspection should occur before any waste is stored in the tank and at least once annually thereafter.

4. For tanks with Level 1 control: Does the tank meet the following three conditions for Level 1 control:
   a. Waste maximum organic vapor pressure less than cutoff for tank design capacity?   Yes___ No___
   b. No heating ≥ temperatures at which vapor pressure is determined (either by knowledge or by measurement)?   Yes___ No___
   c. No waste stabilization in tank?   Yes___ No___
5. Are vapor pressure determinations documented and available?   Yes___ No___
6. For tanks with Level 2/Option 3 controls (fixed roof tank venting through a closed vent system, to a control device that would destroy or reduce at least 95% of vapors):
   a. Is the fixed roof forming a continuous barrier over the entire surface area of the liquid in the tank?   Yes___ No___
   b. Are emissions vented to a control device?   Yes___ No___
   c. Are all openings in the roof not venting to the control device fixed with a closure device?   Yes___ No___
   d. If vapor pressure underneath the fixed roof cover is less than atmospheric pressure when control device is working and the closure device is closed, are any cracks, holes, gaps or other open spaces visible between cover opening and closure device?   Yes___ No___
   e. If vapor pressure below the fixed roof cover equals or exceeds atmospheric pressure when the control device is working, are the cover and closure device designed to operate at NDE?   Yes___ No___
   f. Are the cover and closure devices closed at all times and the vapor headspace vented to a control device except when owner/operator is: (1) performing inspections or (2) conducting maintenance or other normal operations or (3) accessing the tank or (4) removing accumulated sludge and other residues from the bottom of the tank?   Yes___ No___
7. Is monitoring data from the control device and design data documented and available?   Yes___ No___

TABLE A.28
Underground Storage Tank (UST) Checklist

Date: Time:

Inspector's name: Signature:

Note: This is a basic checklist for USTs. There are additional checklists available from EPA and state agencies that address each of the main topical areas presented in this checklist.

Tank ID #

Installation or replacement date

·(**Note:** If tank was installed or replaced after April 11, 2016, it must have secondary containment and interstitial monitoring.)

Release Detection Method for Tanks (check all that apply)

a. Automatic tank gauging (ATG) system _____

b. Interstitial monitoring (with secondary containment) _____

c. Statistical inventory reconciliation (SIR) _____

d. Continuous in-tank leak detection (CITLD) _____

e. Vapor monitoring _____

f. Groundwater monitoring _____

g. Inventory control and tank tightness testing (TTT) _____

[This option allowed only for ten years after the tank was installed. Tank tightness testing (TTT) required every five years.]

h. Manual tank gauging only _____

    (This option allowed only for tanks of 1,000-gallon capacity or less, with specified diameters.)

i. Manual tank gauging and tank tightness testing

    (This option allowed only for tanks of 2,000 gallon capacity or less and only for ten years after tank was installed. TTT required every five years.)

j. Other release detection method (please specify)

Release Detection for Pressurized Piping (check all that apply)

(**Note:** Piping installed or replaced after April 11, 2016, must have secondary containment and interstitial monitoring and have an automatic line leak detector.)

1. Automatic line leak detectors

    a. Automatic flow restrictor _____

    b. Automatic shutoff device _____

    c. Audible or visual alarm _____

2. Other methods

    a. Annual line tightness test _____

    b. Monthly monitoring (includes interstitial monitoring, vapor monitoring, groundwater monitoring, SIR and CITLD). _____

Release Detection for Suction Piping (check all that apply)

(**Note:** Piping installed or replaced after April 11, 2016, must have secondary containment and interstitial monitoring (except for safe suction piping).)

1. Line tightness testing every three years _____

2. Monthly monitoring (includes interstitial monitoring, vapor monitoring, groundwater monitoring and other accepted methods such as SIR and electronic line leak detectors) _____

3. No release detection (safe suction piping system with the following characteristics: only one check valve per line located directly below the dispenser; piping sloping back to the tank; and system must operate under atmospheric pressure.) _____

Spill and Overflow Protection (check all that apply)

1. Spill catchment basin or spill bucket _____

2. Automatic shutoff device _____

3. Overfill alarm _____

4. Ball float valve   (Ball float valves may not be used to meet this requirement when overfill prevention is installed or replaced after October 13, 2015.)

Corrosion Protection for Tanks (check all that apply)

1. Coated and cathodically protected steel _____

2. Noncorrodible material (such as fiberglass-reinforced plastic) _____

3. Steel jacketed or clad with noncorrodible material _____

4. Cathodically protected noncoated steel _____

5. Internally lined tank _____

6. Other method (please specify)

Corrosion Protection for Piping (check all that apply)

1. Coated and cathodically protected steel _____

2. Noncorrodible material (such as fiberglass-reinforced plastic or flexible plastic) _____

3. Cathodically protected noncoated metal _____

4. Other method (please specify)

Appendix B
Toolkit for Evaluating NPDES

This appendix provides a series of checklists intended to assist compliance managers and personnel performing inspections and assessments of some of the more common operations involving storm water and wastewater inspections governed by NPDES. These tools are provided for instructional purpose and guidance only and are not intended to represent an all-inclusive collection of NPDES regulatory requirements. The tools provided in this appendix are adapted from EPA and state environmental agency checklists and organized within this appendix as follows.

Municipal Wastewater:

1. Individual NPDES Permits
2. Water Treatment General NPDES Permit
3. Pretreatment Program for POTWs

Storm Water Pollution Prevention Plan (SWPP) and Monitoring:

4. Construction SWPP
5. Industrial SWPP

TABLE B.1

Individual Municipal or Industrial Wastewater Permit Checklist

Date: _____ Time: _____

Inspector's name: _____ Signature: _____

PERMIT

1. Is a copy of the current permit (Parts I, II and attachments) on-site and available? (40 CFR 122.41)   Yes___ No___

2. If the permit is expired or due to expire within 180 days, has a reapplication package been submitted to appropriate state agency and EPA? (40 CFR 122.21)?   Yes___ No___

RECORDS/REPORTS

3. Are the records and reports maintained by the permittee for at least three years? (40 CFR 122.21(p) and 40 CFR 122.41(j)(2))   Yes___ No___

4. For any permitted parameter monitored more frequently than required by the permit, using approved test methods, are these additional results included in discharge monitoring report (DMR) calculations?   Yes___ No___

5. Are analytical results reported on the facilities benchsheets consistent with data reported on the DMRs?   Yes___ No___

FACILITY SITE REVIEW

6. Is there excessive scum buildup, grease, foam or floating sludge in or on any of the treatment units? (40 CFR 122.41(e))   Yes___ No___

7. Are tank weirs level? (40 CFR 122.41(e))   Yes___ No___

8. Is there any indication of a hydraulic overload? (40 CFR 122.41(e))   Yes___ No___

9. Are there any noxious odors leaving the site? (40 CFR 122.41(e))   Yes___ No___

10. Are there any unsafe conditions (e.g., slicks, faulty guardrails, missing grating, etc.)? (40 CFR 122.41(e))   Yes___ No___

11. Is there any evidence of severe corrosion in any piping or equipment? (40 CFR 122.41(e))   Yes___ No___

12. Are there any breaks or leaks in any chemical feed lines or other piping? (40 CFR 122.41(e))   Yes___ No___

13. Is there any surcharging of influent lines, overflow weirs or other structures? (40 CFR 122.41(e))   Yes___ No___

14. Is there any evidence of septage spills at the septage receiving facility? (40 CFR 122.41(e))   Yes___ No___

15. Are there any unpermitted flows entering the groundwater or surface water from either the wastewater treatment facility or the collection system?   Yes___ No___

16. Is there any evidence of potential spills which can contribute pollutants to any storm drains?   Yes___ No___

17. Is there any dry weather flow in the storm water drainage system within the facility?   Yes___ No___

18. Does the facility have any floor drains that discharge to storm drain system, surface water or groundwater without a permit?   Yes___ No___

(*Continued*)

TABLE B.1 (CONTINUED)
Individual Municipal or Industrial Wastewater Permit Checklist

EFFLUENT/RECEIVING WATER

19. Are there any floating solids, oil sheen, color or foam in the effluent?   Yes___ No___

20. Are there any floating solids, oil sheen, color, foam or a recognizable plume in the
    receiving water?   Yes___ No___

FLOW MEASUREMENT

21. Are influent (if applicable) and effluent flow-measuring device(s) professionally
    calibrated, at least once per year? (40 CFR 122.41(e))   Yes___ No___

    What type of influent meter is used? _____
    What type of effluent meter is used? _____

22. Do personnel check the calibration of the flow-measuring device(s), between the
    annual professional calibrations, at least three times per year?   Yes___ No___

23. Are the results recorded for these additional tests, and are the results within
    10%?   Yes___ No___

24. Are all effluent flow-measuring devices clean and free of debris and deposits? (40
    CFR 122.41(e))   Yes___ No___

25. Are the sides of the flume(s) throat vertical and parallel? (40 CFR 122.41(e))   Yes___ No___

26. Is the effluent weir level? (40 CFR 122.41(e))   Yes___ No___

27. Is there any leakage around any of the flow-measuring devices? (40 CFR
    122.41(e))   Yes___ No___

SELF-MONITORING

28. Are the influent and effluent sampling locations representative of the waste
    stream?   Yes___ No___

29. Are the correct effluent sample types (grab or composite) taken?   Yes___ No___

30. If composite samples are required, are they flow-proportioned?   Yes___ No___

31. Are composite samples cooled to <6°C to properly preserve them during the
    compositing period? (40 CFR 136)   Yes___ No___

32. If the composite sample is cooled with ice or gel packs, is the final composite
    sample temperature measured and recorded to make sure that the cooling is
    sufficient? (40 CFR 122.41(e) and 40 CFR 136)   Yes___ No___

33. If a refrigerator is used for preserving composite samples, is there a thermometer
    in the refrigerator?   Yes___ No___

34. Is this thermometer either checked each time that it is used and results recorded or
    is the final sample temperature measured and the results recorded? (40 CFR
    122.41(e) and 40 CFR 136)   Yes___ No___

35. Are all grab samples cooled with ice, gel packs or refrigerated to <6°C from the
    time of collection until analysis including shipping time, if applicable?   Yes___ No___

36. Are all samples which require preservation properly preserved? (40 CFR
    122.41(e) and 40 CFR 136)   Yes___ No___

37. Are the correct sample containers being used? (40 CFR 122.41(e) and 40 CFR
    136)   Yes___ No___

(Continued)

TABLE B.1 (CONTINUED)
Individual Municipal or Industrial Wastewater Permit Checklist

38. Are all the sampling equipment and glassware cleaned before being used? (40 CFR 122.41(e) and 40 CFR 136)   Yes___ No___

39. Does the facility's permit require any metals sampling?   Yes___ No___

40. If yes to 40, does the facility acid wash the sampling containers prior to sample collection as required by the approved analytical methods as required by the facility's permit?   Yes___ No___

LABORATORY

41. Has a written laboratory QA/QC manual been updated by the facility and approved by State regulatory agency in the last five years? (40 CFR 122.41(e))   Yes___ No___

42. Is the QA/QC manual being used by facility personnel?   Yes___ No___

43. Does the facility have a copy of the EPA-approved analytical methods for each of the analyses performed at the facility?   Yes___ No___

44. Are the correct analytical testing procedures used and holding times met? (40 CFR 136)   Yes___ No___

45. Are laboratory method detection limits for all parameters tested less than the permit limits?   Yes___ No___

46. With each batch of samples analyzed, are quality control standards, sample duplicates, spikes and blanks conducted? (40 CFR 136)   Yes___ No___

47. If using alternative analytical procedures, have they been approved by EPA? (40 CFR 136)   Yes___ No___

48. Are all laboratory instruments and equipment being calibrated and maintained on the periodic basis specified in the Part 136 Analytical Method or in the QA/QC Manual? (40 CFR 122.41(e) and 40 CFR 136)   Yes___ No___

49. Are the thermometers annually checked for calibration using a NIST-certified thermometer or do you purchase new NIST-certified thermometers yearly? (40 CFR 122.41(e))   Yes___ No___

50. Are the reagents and standards being used expired? (40 CFR 122.41(e))   Yes___ No___

51. Is proper laboratory-grade laboratory-pure water available for specific analyses? (40 CFR 122.41(e) and 40 CFR 136)   Yes___ No___

52. Are laboratory safety devices (eyewash and shower, fume hood, proper labeling and storage, pipette suction bulbs) available? (Recommendation only)   Yes___ No___

53. Are reagents and solvents used for the analyses properly stored? (40 CFR 122.41(e), 40 CFR 136)   Yes___ No___

54. Are the correct lab formulas to calculate final results? (40 CFR 136)   Yes___ No___

OPERATIONS AND MAINTENANCE

55. Are all treatment units operable? (40 CFR 122.41(e))   Yes___ No___

56. Does the wastewater treatment facility have an alarm system for all essential equipment? (40 CFR 122.41(e))   Yes___ No___

57. Are the facility alarm systems checked? How often? When were the alarm systems last checked? (40 CFR 122.41(e))   Yes___ No___

(Continued)

TABLE B.1 (CONTINUED)
Individual Municipal or Industrial Wastewater Permit Checklist

58. Are alarms sent to qualified personnel who can respond immediately to remedy the problem? (40 CFR 122.41(e))   Yes___ No___

59. Are routine and preventive maintenance scheduled performed and recorded? (40 CFR 122.41(e))   Yes___ No___

60. Does the facility maintain written procedures for responding to emergencies such as power failures, floods, fires and other natural disasters? (40 CFR 122.41(e))   Yes___ No___

61. Does the facility maintain a written list of contacts for emergencies? (40 CFR 122.41(e))   Yes___ No___

62. Is a logbook kept which documents all plant activities on a daily basis? (40 CFR 122.41(e) and 40 CFR 122.41(j)(2))   Yes___ No___

63. Does the facility maintain an inventory of spare parts, either at the facility or close by, sufficient to keep all of its treatment units operational? (40 CFR 122.41(e))   Yes___ No___

64. Does the facility have standby power for all treatment units? (40 CFR 122.41(e))   Yes___ No___

65. Is the standby power regularly exercised under load? (40 CFR 122.41(e))   Yes___ No___

66. Have there been any backups or overflows in the sanitary sewer collection system, including pump stations, manholes and piping in the last two years?   Yes___ No___

67. If yes to #66, are these overflows reported to the appropriate regulatory authority within 24 hours verbally and followed up with a letter in five days?   Yes___ No___

68. If yes to #66, have any of these overflows impacted surface water?   Yes___ No___

69. Does the storm water collection system for the municipality have any dry weather flows?   Yes___ No___

70. Does the facility have up-to-date maps/schematics of all storm water outfalls?   Yes___ No___

TABLE B.2
Water Treatment General NPDES Permit Checklist

Date: _____ Time: _____

Inspector's name: _____ Signature: _____

PERMIT
1. Is a copy of the current permit (Parts I, II and attachments) on-site and available? (40 CFR 122.41)   Yes___ No___

2. If the permit is expired or due to expire within 180 days, has a reapplication package been submitted to appropriate state agency and EPA? (40 CFR 122.21)?   Yes___ No___

RECORDS/REPORTS
3. Are the records and reports maintained by the permittee for at least three years? (40 CFR 122.21(p) and 40 CFR 122.41(j)(2))   Yes___ No___

4. For any permitted parameter monitored more frequently than required by the permit, using approved test methods, are these additional results included in discharge monitoring report (DMR) calculations?   Yes___ No___

5. Are analytical results reported on the facilities benchsheets consistent with data reported on the DMRs?   Yes___ No___

FACILITY SITE REVIEW
6. Is there excessive scum buildup, grease, foam or floating sludge in or on any of the treatment units? (40 CFR 122.41(e))   Yes___ No___

7. Are tank weirs level? (40 CFR 122.41(e))   Yes___ No___

8. Is there any indication of a hydraulic overload? (40 CFR 122.41(e))   Yes___ No___

9. Are there any noxious odors leaving the site? (40 CFR 122.41(e))   Yes___ No___

10. Are there any unsafe conditions (e.g., slicks, faulty guardrails, missing grating, etc.)? (40 CFR 122.41(e))   Yes___ No___

11. Is there any evidence of severe corrosion in any piping or equipment? (40 CFR 122.41(e))   Yes___ No___

12. Are there any breaks or leaks in any chemical feed lines or other piping? (40 CFR 122.41(e))   Yes___ No___

13. Are there any unpermitted flows entering the groundwater or surface water from either the wastewater treatment facility or the collection system?   Yes___ No___

14. Is there any evidence of potential spills which can contribute pollutants to any storm drains?   Yes___ No___

15. Is there any dry weather flow in the storm water drainage system within the facility?   Yes___ No___

16. Does the facility have any floor drains that discharge to storm drain system, surface water or groundwater without a permit?   Yes___ No___

FLOW MEASUREMENT

17. Is the effluent flow-measuring device(s) professionally calibrated in accordance with the manufacturer's recommendations? (40 CFR 122.41(e))   Yes___ No___

18. Do personnel check the calibration between these professional calibrations?   Yes___ No___

(Continued)

TABLE B.2 (CONTINUED)
Water Treatment General NPDES Permit Checklist

19. Are all effluent flow-measuring devices clean and free of debris and deposits? (40 CFR 122.41(e))   Yes___ No___

20. Is the effluent weir level? (40 CFR 122.41(e))   Yes___ No___

21. Is there any leakage around any of the flow-measuring devices? (40 CFR 122.41(e))   Yes___ No___

SELF-MONITORING

22. Is the effluent sampling location representative of the waste stream?   Yes___ No___

23. Are the correct effluent sample types (grab or composite) taken?   Yes___ No___

24. Is the influent sampling location representative of the waste stream?   Yes___ No___

25. Are all grab samples cooled with ice, gel packs or refrigerated to <6°C from the time of collection until analysis including shipping time, if applicable?   Yes___ No___

26. Are all samples which require preservation properly preserved? (40 CFR 122.41(e) and 40 CFR 136)   Yes___ No___

27. Are the correct sample containers being used? (40 CFR 122.41(e) and 40 CFR 136)   Yes___ No___

28. Are all the sampling equipment and glassware cleaned before being used? (40 CFR 122.41(e) and 40 CFR 136)   Yes___ No___

29. Does the facility's permit require any metals sampling?   Yes___ No___

30. If yes to 40, does the facility acid wash the sampling containers prior to sample collection as required by the approved analytical methods or as required by the facility's permit?   Yes___ No___

LABORATORY

31. Has a written laboratory QA/QC manual been updated by the facility and approved by State regulatory agency in the last five years? (40 CFR 122.41(e))   Yes___ No___

32. Is the QA/QC manual being used by facility personnel?   Yes___ No___

33. Does the facility have a copy of the EPA-approved analytical methods for each of the analyses performed at the facility?   Yes___ No___

34. Are the correct analytical testing procedures used and holding times met? (40 CFR 136)   Yes___ No___

35. Are laboratory method detection limits for all parameters tested less than the permit limits?   Yes___ No___

36. With each batch of samples analyzed, are quality control standards, sample duplicates, spikes and blanks conducted? (40 CFR 136)   Yes___ No___

37. If using alternative analytical procedures, have they been approved by EPA? (40 CFR 136)   Yes___ No___

(Continued)

TABLE B.2 (CONTINUED)
Water Treatment General NPDES Permit Checklist

38. Are all laboratory instruments and equipment being calibrated and maintained on the periodic basis specified in the Part 136 Analytical Method or in the QA/QC Manual? (40 CFR 122.41(e) and 40 CFR 136)   Yes___ No___

39. Are the thermometers annually checked for calibration using a NIST-certified thermometer or do you purchase new NIST-certified thermometers yearly? (40 CFR 122.41(e))   Yes___ No___

40. Are the reagents and standards being used expired? (40 CFR 122.41(e))   Yes___ No___

41. Is proper laboratory-grade laboratory-pure water available for specific analyses? (40 CFR 122.41(e) and 40 CFR 136)   Yes___ No___

42. Are laboratory safety devices (eyewash and shower, fume hood, proper labeling and storage, pipette suction bulbs) available? (Recommendation only)   Yes___ No___

43. Are reagents and solvents used for the analyses properly stored? (40 CFR 122.41(e) and40 CFR 136)   Yes___ No___

44. Are the correct lab formulas to calculate final results? (40 CFR 136)   Yes___ No___

OPERATIONS AND MAINTENANCE
45. Are all treatment units operable? (40 CFR 122.41(e))   Yes___ No___

46. Does the wastewater treatment facility have an alarm system for all essential equipment? (40 CFR 122.41(e))   Yes___ No___

47. Are the facility alarm systems checked? How often? When were the alarm systems last checked? (40 CFR 122.41(e))   Yes___ No___

48. Are alarms sent to qualified personnel who can respond immediately to remedy theproblem? (40 CFR 122.41(e))   Yes___ No___

49. Are routine and preventive maintenance scheduled performed and recorded? (40 CFR 122.41(e))   Yes___ No___

50. Does the facility maintain written procedures for responding to emergencies such as power failures, floods, fires and other natural disasters? (40 CFR 122.41(e))   Yes___ No___

51. Does the facility maintain a written list of contacts for emergencies? (40 CFR 122.41(e))   Yes___ No___

52. Is a logbook kept which documents all plant activities on a daily basis? (40 CFR 122.41(e) and 40 CFR 122.41(j)(2))   Yes___ No___

53. Does the facility maintain an inventory of spare parts, either at the facility or close by, sufficient to keep all of its treatment units operational? (40 CFR 122.41(e))   Yes___ No___

54. Does the facility have standby power for all treatment units? (40 CFR 122.41(e))   Yes___ No___

55. Is the standby power regularly exercised under load? (40 CFR 122.41(e))   Yes___ No___

   Does the facility have up-to-date maps/schematics of all storm water outfalls?   Yes___ No___

TABLE B.3
POTW Pretreatment Program Checklist

Date: _____ Time: _____

Inspector's name: _____ Signature: _____

1. Have specific limits been developed for the POTW to ensure compliance with 40
   CFR 403.5(a), 403.5(b), 40 CFR 403.5(c) and 403.8(f)(4)?   Yes___ No___

2. Does the pretreatment program continue to develop these limits as necessary and
   effectively enforce such limits?   Yes___ No___

3. Is the total design flow greater than 5 million gallons per day (mgd)?   Yes___ No___

4. Are pollutants received from industrial users which pass through or interfere with
   the operation of the POTW?   Yes___ No___

5. If yes to #4, is there a POTW pretreatment program which meets specific criteria
   (40 CFR 403.8(a) and 403.8(f)(2))?   Yes___ No___

6. Does the pretreatment program includes procedures which enables the POTW to:

   a. Identify and locate all possible industrial users which might be subject to the
      POTW pretreatment program?   Yes___ No___

   b. Identify the character and volume of pollutants contributed to the POTW by the
      identified industrial users?   Yes___ No___

   c. Notify identified industrial users of applicable pretreatment standards and any
      applicable requirements under Sections 204(b) and 405 of the CWA and
      Subtitles C and D of RCRA?   Yes___ No___

   d. Receive and analyze self-monitoring reports and other notices submitted by
      industrial users in accordance with the self-monitoring requirements?   Yes___ No___

   e. Randomly sample and analyze the effluent from industrial users and conduct
      surveillance activities in order to identify, independent of information supplied
      by industrial users, occasional and continuing noncompliance with pretreatment
      standards?   Yes___ No___

   f. Inspect and sample the effluent from each significant industrial user at least
      once a year?   Yes___ No___

   g. Evaluate, at least once every two years, whether each significant industrial user
      needs a plan to control slug discharges?   Yes___ No___

   h. Investigate instances of noncompliance with pretreatment standards and
      requirements, as indicated in the required reports and notices, or indicated by
      analysis, inspection and surveillance activities?   Yes___ No___

   i. Perform sample taking and analysis and the collection of other information
      with sufficient care to produce evidence admissible in enforcement proceedings
      or in judicial actions?   Yes___ No___

   j. Comply with the public participation requirements of 40 CFR 25 in the
      enforcement of national pretreatment standards?   Yes___ No___

7. Do procedures include provisions for at least annual public notification in the
   largest daily newspaper published in the local municipality, of industrial users
   which, at any time during the previous 12 months, were in significant
   noncompliance with applicable pretreatment requirements?   Yes___ No___

8. Does the request for approval of a POTW pretreatment program submitted to
   approval authority include a program description containing the following information:

   a. A statement from the City Solicitor or a city official acting in a comparable
      capacity (or independent legal counsel) that the POTW has authority adequate
      to carry out the programs described in 40 CFR 403.8?   Yes___ No___

(Continued)

TABLE B.3 (CONTINUED)
POTW Pretreatment Program Checklist

　b. Does the statement identify the provision of the legal authority under 40 CFR 403.8(f)(1) which provides the basis for each procedure under 40 CFR 403.8(f)(2)? Yes___ No___

　c. Identify the manner in which the POTW will implement the program requirements set forth in 40 CFR 403.8, including the means by which pretreatment standards will be applied to individual industrial users (e.g., by order, permit, ordinance, etc.)? Yes___ No___

　d. Identify how the POTW intends to ensure compliance with pretreatment standards and requirements, and to enforce them in the event of noncompliance by industrial users a copy of any statutes, ordinances, regulations, agreements or other authorities relied upon by the POTW for its administration of the program, including a statement reflecting the endorsement or approval of the local boards or bodies responsible for supervising and/or funding the POTW pretreatment program if approved? Yes___ No___

　e. A brief description (including organization charts) of the POTW organization which will administer the pretreatment program. If more than one agency is responsible for administration of the program, the responsible agencies should be identified, their respective responsibilities delineated and their procedures for coordination set forth? Yes___ No___

　f. A description of the funding levels and full-and part-time manpower available to implement the program? Yes___ No___

9. Are there sufficient resources of qualified personnel to carry out the POTW pretreatment program (40 CFR 403.8(f)(3))? Yes___ No___

10. Has an enforcement response plan been developed and implemented? (40 CFR 403.8(f)(5)) Yes___ No___

11. Has a list of significant industrial users been prepared and being maintained? (40 CFR 403.8(f)(6))? Yes___ No___

12. Does the annual report provided to the approval authority contain the following information? (40 CFR 403.12(i))

　a. An updated list of the POTW's industrial users, including their names and addresses, or a list of deletions and additions keyed to a previously submitted list? Yes___ No___

　b. Does the information include

　　i.　A brief explanation of each deletion? Yes___ No___
　　ii.　Which Industrial Users are subject to categorical pretreatment standards and which standards are applicable to each industrial user? Yes___ No___
　　iii.　Which Industrial Users are subject to local standards that are more stringent than the categorical pretreatment standards? Yes___ No___
　　iv.　The Industrial Users that are subject only to local Requirements? Yes___ No___

　c. A summary of the status of industrial user compliance over the reporting period? Yes___ No___

　d. A summary of compliance and enforcement activities (including inspections) conducted by the POTW during the reporting period? Yes___ No___

　e. A summary of changes to the POTW's pretreatment program that have not been previously reported to the approval authority? Yes___ No___

　f. Any other relevant information requested by the approval authority? Yes___ No___

(Continued)

TABLE B.3 (CONTINUED)
POTW Pretreatment Program Checklist

13. Are records kept for all monitoring activities? (40 CFR 403.12(o))   Yes___ No___

14. Do the records for all samples include:

    a. The date, exact place, methods and time of sampling and the names of the
       person or persons taking the samples?   Yes___ No___

    b. The dates analyses were performed?   Yes___ No___

    c. Who performed analyses?   Yes___ No___

    d. The analytical techniques, methods used?   Yes___ No___

    e. The results of the analyses?   Yes___ No___

15. Are records kept for three years and signed and certified by the equivalent of a
    responsible corporate officer?   Yes___ No___

16. Is facility required to collect whole effluent toxicity (WET) data which must meet
    specific requirements? (40 CFR 122.21(j))   Yes___ No___

17. Have the results of valid whole effluent biological toxicity testing been provided to
    the approval authority?   Yes___ No___

    Has a written technical evaluation of the need to revise local limits under 40 CFR
    403.5(c)(1) been submitted to the approval authority?   Yes___ No___

TABLE B.4
Construction Storm Water Checklist

Date: _____ Time: _____

Inspector's name: _____ Signature: _____

1. Has a construction SWPP been developed for projects that disturb 1 or more acres of land?   Yes___ No___

2. Have measures or controls been incorporated into the SWPPP that are consistent with the assumptions and requirements of the TMDL?   Yes___ No___

3. Are measure implemented (i.e., BMPs) to protect listed threatened or endangered species or critical habitat?   Yes___ No___

4. Are measures implemented to protect historic properties and places?   Yes___ No___

5. Is a definable area marked on the SWPPP being stabilized?   Yes___ No___

6. Have the activities negatively impacted another party's pollution controls?   Yes___ No___

7. Are the disturbed areas stabilized (sodded or covered by other means) where construction activities have temporarily or permanently ceased?   Yes___ No___

8. Are the appropriate sediment and erosion control measures for the drainage area of the site is implemented as described in the SWPPP?   Yes___ No___

9. Is the placement of structural measures in the SWPPP consistent with the site?   Yes___ No___

10. Verify on-site through inspection of outfalls that no solid/building materials are discharged unless authorized by a permit.   Yes___ No___

11. Is the accumulation of sediment kept inside the construction entrance or perimeter of the construction site?   Yes___ No___

12. Do the construction and waste materials stored on-site match those noted in the SWPPP?   Yes___ No___

13. Have the exposure minimization practices discussed in the SWPPP have been implemented?   Yes___ No___

14. Have the spill prevention and response practices being implemented (there are materials provided to employees, trainings, etc.)?   Yes___ No___

15. Have velocity dissipation devices been installed?   Yes___ No___

16. Verify there is no evidence of erosion at outlet?   Yes___ No___

17. Has high-velocity flow not misplaced rocks?   Yes___ No___

18. Are the size and number of rocks adequate to avoid soil erosion?   Yes___ No___

19. There is no evidence of additional sedimentation in receiving water attributable to the site?   Yes___ No___

20. Are non-storm water discharges eliminated or reduced to the extent feasible?   Yes___ No___

21. Are the control measures, including erosion and sediment controls, identified in the SWPPP installed and operating effectively?   Yes___ No___

22. Has any control measure that has been used inappropriately or correctly been replaced or modified?   Yes___ No___

(Continued)

TABLE B.4 (CONTINUED)
Construction Storm Water Checklist

23. Verify on-site that changes to existing BMPs or additional BMPs have been implemented/added? Yes___ No___

24. Sediment accumulated does not exceed 50% of capacity of sediment traps and/or sediment ponds? Yes___ No___

25. If inspection frequency has been modified,

    a. Has the entire site been temporarily stabilized? Yes___ No___

    b. Runoff is unlikely due to winter conditions (e.g., site is covered with snow, ice or the ground is frozen)? Yes___ No___

    c. Construction is occurring during seasonal arid periods in arid and semiarid areas? Yes___ No___

TABLE B.5
Industrial Storm Water Checklist

Date: _____ Time: _____

Inspector's name: _____ Signature: _____

1. Have the following areas been inspected at the facility?

   a. Bulk storage areas (tanks, drums, fuels, pallets, etc.)　　　　Yes___ No___

   Inspection results: _____
   Actions required: _____

   b. Waste disposal areas　　　　Yes___ No___

   Inspection results: _____
   Actions required: _____

   c. Maintenance areas　　　　Yes___ No___

   Inspection results: _____
   Actions required: _____

   d. Loading/Unloading areas　　　　Yes___ No___

   Inspection results: _____
   Actions required: _____

   e. Raw material, product, by-product and final storage areas　　　　Yes___ No___

   Inspection results: _____
   Actions required: _____

   f. Other areas:　　　　Yes___ No___

   Inspection results: _____
   Actions required: _____

2. Are structural and nonstructural Best Management Practices (BMPs) functioning
   properly?　　　　Yes___ No___

   Describe any identified issues: _____
   _____
   _____

3. Was runoff evaluated during a storm event?　　　　Yes___ No___

   Describe discharge(s) _____
   _____

4. Was facility inspected for oil sheen?　　　　Yes___ No___

   Results of inspection: _____
   _____

5. Is sweeping being conducted at the facility?　　　　Yes___ No___

6. Frequency of sweeping

7. Does current SWPP adequately describe site conditions?　　　　Yes___ No___

Appendix C
Evaluating Air Emissions

This appendix provides a series of checklists intended to assist compliance managers and personnel performing inspections and assessments of some of the more common operations involving air emission inspections governed by the CAA. These tools are provided for instructional purpose and guidance only and are not intended to represent an all-inclusive collection of CAA regulatory requirements. The tools provided in this appendix are adapted from EPA and state environmental agency checklists. It is noted that the majority of compliance inspections of permitted sources are an assessment of adherence to the requirements and conditions stipulated in the permit for the specific permitted source. The example checklists provided in this appendix are as follows:

CAA Compliance Inspections:

1. Emergency Standby/ Portable Diesel Building Generator
2. Emergency Boiler
3. Asbestos Notifications: Demolition
4. Asbestos Notifications: Renovation

TABLE C.1
Emergency Standby/Portable Diesel Generator Checklist

Date: _____ Time: _____
Inspector's name: _____ Signature: _____

1. Nonresettable hour meter and hour meter are present and functioning?   Yes___ No___
2. List engine use per year:
3. Are emergency and maintenance test records maintained and available?   Yes___ No___
4. Are the records of engine use consistent with allowable hours stated in the permit
   condition?   Yes___ No___
5. Do the Serial Number/Horse Power/Engine Tier Number (based on EPA emission
   test) of the engine match those listed on the permit?   Yes___ No___
6. Is a copy of the permit maintained in the files and readily available?   Yes___ No___
7. Has the engine been started and a visible emission verification for Ringlemann 1%
   or 20% Visibility test performed?   Yes___ No___
8. Is the person performing the visibility test certified to conduct visible emission
   verification?   Yes___ No___
9. Are the permit conditions for sulfur content of the diesel fuel being met?   Yes___ No___
10. If yes to #9, do diesel fuel records identify the fuel supplier and amount of sulfur
    content in the fuel?   Yes___ No___

TABLE C.2
Emergency Boiler Checklist

Date: _____ Time: _____

Inspector's name: _____ Signature: _____

1. Is the Boiler BTU per hour input rating ≤ 2 million BTU/hr.?   Yes___ No___
2. If yes to #1, the boiler is exempt from permitting. Evaluation recorded?   Yes___ No___
3. Is the boiler rating >2 million to 10 million BTU/hr.?   Yes___ No___
4. If yes to #3, has boiler been register with regulatory authority?   Yes___ No___
5. Is boiler rating >10 million BTU/hr.?   Yes___ No___
6. Boiler requires a permit, has permit been obtained?   Yes___ No___
7. Do the serial number/model number/installation date of the boiler burner match
   information contained in the records?   Yes___ No___
8. Is the boiler >10 years old?   Yes___ No___
9. If yes to #8, has the boiler burner been changed to low NOx?   Yes___ No___
10. Are annual source test records of the boiler emission available?   Yes___ No___
11. Have they been verified against the amount of NOx and CO requirements specified
    in the rules and regulations based on the BTU/hr. heat input rating?   Yes___ No___
12. Are annual calibration and certification records for source test equipment
    maintained and accessible?   Yes___ No___
13. Is a 1-inch gap maintained between the sections of the boiler pipe insulation
    (except around valves, monitoring devices and safety devices for the steam supply
    line coming out of the boiler and not for the steam return line to the
    boiler)?   Yes___ No___
14. Is a copy of the current permit maintained and available?   Yes___ No___

TABLE C.3
Notification for Asbestos Demolition Projects Checklist

Date: _____ Time: _____

Inspector's name: _____ Signature: _____

Based on NESHAP regulations:

1. Has notification been provided to the regulatory authority at least ten days prior to
   the actual start of work?   Yes___ No___
2. Have required fees been paid?   Yes___ No___
3. Has approval of the demolition been received along with demolition permit?   Yes___ No___
4. Are buildings and or locations properly identified?   Yes___ No___
5. Is the start date of the actual demolition consistent with permit?   Yes___ No___
6. Is the end date of the actual demolition work consistent with the permit?   Yes___ No___
7. Is the method of demolition consistent with permit?   Yes___ No___
8. Are all personnel conducting the work current with asbestos annual certification? Yes___ No___

TABLE C.4
Notification for Asbestos Renovation Projects Checklist

noindent

Date: Time:

Inspector's name: Signature:

Based on NESHAP regulations:

1. Has notification been provided to the regulatory authority at least ten days prior to the actual start of work? — Yes___ No___
2. Have required fees been paid? — Yes___ No___
3. Has approval of the renovation been received along with demolition permit? — Yes___ No___
4. Are buildings and or locations properly identified? — Yes___ No___
5. Is the start date of the actual demolition consistent with permit? — Yes___ No___
6. Is the end date of the actual renovation work consistent with the permit? — Yes___ No___
7. Is the method of renovation consistent with permit? — Yes___ No___
8. Are all personnel conducting the work current with asbestos annual certification? — Yes___ No___
9. Has the amount of friable (at or greater than 1%) asbestos to be stripped be identified? — Yes___ No___
10. Are the aspects of the friable asbestos characterization (sampling, method used and composite characterization) properly identified? — Yes___ No___
11. Is the method of removal, waste packaging, labeling and disposal identified (mechanical methods of removal may require an exemption from the regulatory authority)? — Yes___ No___
12. Are containment methods identified and consistent with requirements? — Yes___ No___
13. Are the correct waste regulatory requirements identified? — Yes___ No___

APPENDIX D
Toolkit for Evaluating EPCRA

This appendix provides a series of checklists intended to assist compliance managers and personnel performing inspections and assessments of some of the more common operations involving planning, reporting and record-keeping governed by EPCRA. These tools are provided for instructional purpose and guidance only and are not intended to represent an all-inclusive collection of EPCRA regulatory requirements. The tools provided in this appendix are adapted from EPA and organized within this appendix as follows:

EPCRA:

1. Planning
2. Release Notification/Reporting
3. Record-keeping

TABLE D.1
EPCRA Planning Checklist

Date: _____ Time: _____
Inspector's name: _____ Signature:_____

1. Does the facility have quantities of extremely hazardous substances equal to or greater than the threshold limitations which require following specific emergency planning procedures? (40 CFR 355.30 and 355 Appendix A)   Yes___ No___

2. Has the state emergency response commission been notified that the facility is subject to the emergency planning requirements within 60 days after first becoming subject to these requirements?   Yes___ No___

3. Has the facility designated a representative who participates in the local emergency planning process as a facility emergency response coordinator?   Yes___ No___

4. Has the facility notified the local emergency planning committee, or governor if there is no committee, of the facility representative within 30 days after establishment of a local emergency planning committee?   Yes___ No___

5. Has the local emergency planning committee been informed of any changes occurring at the facility that may be relevant to emergency planning?   Yes___ No___

6. Upon request of the local emergency planning committee, has the facility promptly provided to the committee any information necessary for development or implementation of the local emergency plan?   Yes___ No___

7. Is the facility contingency plan compatible with the contingency plan developed by the local emergency planning committee?   Yes___ No___

8. Does the facility contingency plan consider how local emergency response officials will likely respond to a chemical release?   Yes___ No___

TABLE D.2
EPCRA Release/Notification Reporting Checklist

Date: _____ Time: _____
Inspector's name: _____ Signature: _____

1. Are facility personnel cognizant of the fact that when there is a release of a reportable quantity (RQ) of any extremely hazardous substance or CERCLA hazardous substance emergency release notification is required (40 CFR 355.40 and 355 Appendices A and B)?   Yes___ No___

2. Have any releases occurred in excess of reportable quantities?   Yes___ No___

3. If yes to #2, were the following notifications made?

   a. Community emergency coordinator for the local emergency planning committee of any area likely to be affected by the release?   Yes___ No___

   b. State emergency response commission of any state likely to be affected by the release?

                                                                          Yes___ No___

   c. Local emergency response personnel if there is no local emergency planning committee?

                                                                          Yes___ No___

4. Did notifications contain the following information:

   a. The chemical name or identity of any substance involved in the release?   Yes___ No___

   b. An indication of whether the substance is an extremely hazardous substance?   Yes___ No___

   c. An estimate of the quantity of any such substance that was released into the environment?

                                                                          Yes___ No___

   d. The time and duration of the release?   Yes___ No___

   e. The medium or media into which the release occurred?   Yes___ No___

   f. Any known or anticipated acute or chronic health risks associated with the emergency, and, where appropriate, advice regarding medical attention necessary for exposed individuals?   Yes___ No___

   g. Proper precautions to take as a result of the release, including evacuation (unless such information is readily available to the community emergency coordination because of the local emergency plan)?   Yes___ No___

   h. The names and telephone numbers of the person or persons to be contacted for further information?   Yes___ No___

5. Was a written follow-up emergency notification produced after the immediate verbal notification which contains the same information detailed in the verbal notice (outlined above)? plus items a–c below: Yes___ No___

   a. Actions taken to respond to and contain the release?   Yes___ No___

   b. Any known or anticipated acute or chronic health risks associated with the release?   Yes___ No___

   c. Advice regarding medical attention necessary for exposed individuals?   Yes___ No___

6. Have releases in excess of or equal to the RQ of listed and unlisted hazardous substances been reported to the NRC immediately (40 CFR 302.5 through 302.6)?   Yes___ No___

7. Are releases of mixtures or solutions (including hazardous waste streams) of hazardous substances, except for radionuclides, reported when either of the following occur:

   a. The quantity of all hazardous constituents of the mixture or solution is known and a reportable quantity or more of any hazardous constituent is released?   Yes___ No___

   b. The quantity of one or more of the hazardous constituents of the mixture or solution is unknown and the total amount of the mixture or solution released equals or exceeds the reportable quantity for the hazardous constituent with the lowest reportable quantity?   Yes___ No___

(Continued)

TABLE D.2 (CONTINUED)
EPCRA Release/Notification Reporting Checklist

8. Have there been any continuous and stable releases of hazardous substances that qualify for reduced reporting options (40 CFR 302.8)?   Yes___ No___

9. For yes to #8, have the following notifications been made:
   a. Initial telephone notification?   Yes___ No___
   b. Initial written notification within 30 days of the initial telephone notification?   Yes___ No___
   c. Follow-up notification within 30 days of the first anniversary date of the initial written notification?   Yes___ No___
   d. Notification of changes in the composition or source of the release or other information submitted in the initial written notification?   Yes___ No___
   e. Information submitted in the follow-up notification when there is an increase in the quantity of the hazardous substances in any 24-h period that represents a statistically significant increase?   Yes___ No___

10. Prior to making an initial telephone notification of a continuous release, did the person in charge establish a sound basis for qualifying the release for reporting by one of the following:
    a. Release data, engineering estimates, knowledge of operating procedures or best professional judgment to establish the continuity and stability of the release?   Yes___ No___
    b. Reporting the release to the NRC for a period sufficient to establish the continuity and stability of the release or when a basis has been established to qualify the release for reduced reporting, initial notification to the NRC is made by telephone?   Yes___ No___

11. Was the notification identified as an initial continuous release notification report and include the following information: the name(s) and location(s) of the facility or vessel; the name(s) and identity(ies) of the hazardous substances being released?   Yes___ No___

12. Was the initial written notification of a continuous release made to the appropriate U.S. EPA Regional Office within 30 days of the initial telephone notification to the NRC?   Yes___ No___

13. Does the initial written notification include, for each release for which reduced reporting as a continuous release is claimed, the following information:
    a. The name of the facility or vessel; the location, including the latitude and longitude; the case number assigned by the NRC or the U.S. EPA; the Dun and Bradstreet number of the facility, if available; the port of registration of the vessel; the name and telephone number of the person in charge of the facility or vessel?   Yes___ No___
    b. The population density within a 1-mile radius of the facility or vessel, described in terms of the following ranges: 0–50 persons, 51–100 persons, 101–500 persons, 501–1,000 persons, more than 1,000 persons?   Yes___ No___
    c. The identity and location of sensitive populations and ecosystems within a 1-mile radius of the facility or vessel (e.g., elementary schools, hospitals, retirement communities or wetlands)?   Yes___ No___

14. Is the following information provided for each hazardous substance release claimed to qualify for reporting under CERCLA Section 103(f)(2):
    a. The name/identity of the hazardous substance?   Yes___ No___
    b. The CAS Registry Number for the substance (if available); and, if the substance being released is a mixture, the components of the mixture and their approximate concentrations and quantities, by weight?   Yes___ No___
    c. The upper and lower bounds of the normal range of the release (in pounds or kilograms) over the previous year?   Yes___ No___
    d. The source(s) of the release (e.g., valves, pump seals, storage tank vents, stacks). If the release is from a stack, the stack height (in feet or meters)?   Yes___ No___
    e. The frequency of the release and the fraction of the release from each release source and the specific period over which it occurs?   Yes___ No___
    f. A brief statement describing the basis for stating that the release is continuous and stable in quantity and rate?   Yes___ No___
    g. An estimate of the total annual amount that was released in the previous year (in pounds or kilograms)?   Yes___ No___
    h. The environmental medium affected by the release, such as the name of the surface water body; the stream order or average flowrate (in cubic feet/second) and designated use; the surface area (in acres) and average depth (in feet or meters) of the lake; the location of public water supply wells within 2 miles if on or underground?   Yes___ No___
    i. A signed statement that the hazardous substance release described is continuous and stable in quantity and rate and that all reported information is accurate and current to the best knowledge of the person in charge?   Yes___ No___

(Continued)

TABLE D.2 (CONTINUED)
EPCRA Release/Notification Reporting Checklist

15. Has notification of any statistically significant increase in a release is made to the NRC as soon as there is knowledge of the release?   Yes___ No___

16. Is every hazardous substance release evaluated annually to determine if changes have occurred in the information submitted in the initial written notification, the follow-up notification and/or in a previous change notification?   Yes___ No___

17. Are supporting materials kept on file for a period of one year and do they substantiate the reported normal range of releases, the basis for stating that the release is continuous and stable in quantity and rate and the other information in the initial written report, the follow-up report and the annual evaluations?   Yes___ No___

18. Is the facility required to prepare or have available a MSDS for a hazardous chemical under OSHA?   Yes___ No___
    If yes to #18, the facility must meet specific MSDS reporting requirements for planning purposes (40 CFR 370.20(a) through 370.21(a), 370.20(c) and 370.28).

19. Are MSDSs (or a list as described below) submitted to the emergency response commission, the local emergency planning committee and the fire department with jurisdiction over the facility for each hazardous chemical present according to the following thresholds:
    a. For all extremely hazardous substances present in amounts greater than or equal to 500 lb (227 kg, approximately 55 gal) or the threshold planning quantity, whichever is lower?   Yes___ No___
    b. For gasoline (all grades combined) in amounts greater than or equal to 75,000 gal (or approximately 283,900 L) when the gasoline is in tanks entirely underground at a retail gas station that was in compliance during the preceding calendar year with all applicable UST regulations (40 CFR Part 280 or requirements of the state UST program approved by U.S. EPA under 40 CFR Part 281)?   Yes___ No___
    c. For diesel fuel (all grades combined) in amounts greater than or equal to 100,000 gal (or approximately 378,500 L) when the diesel is in tanks entirely underground at a retail gas station that was in compliance during the preceding calendar year with all applicable UST regulations (40 CFR Part 280 or requirements of the state UST program approved by U.S. EPA under 40 CFR Part 281)?   Yes___ No___
    d. For all other hazardous chemicals present at any one time in amounts equal to or greater than 10,000 lb (4,540 kg)?   Yes___ No___

20. If the facility has not submitted MSDSs, have the following have been submitted:
    a. A list of hazardous chemicals for which the MSDS is required, grouped by hazard category (see Key Terms and Definitions section of this document for a definition of Hazard Category)?   Yes___ No___
    b. The chemical or common name of each hazardous chemical as provided on the MSDS?   Yes___ No___
    c. Any hazardous component of each hazardous chemical as provided on the MSDS unless reported as a mixture (see 40 CFR 370.28(a)(2))?   Yes___ No___

21. Are revised MSDSs provided to the local emergency planning committee, emergency response commission and fire department within three months after the discovery of significant new information concerning the hazardous chemical for which the MSDSs were submitted?   Yes___ No___

22. Are the Tier I (or Tier II) Hazardous Chemical Inventory forms submitted annually to the local emergency planning committee, the emergency response commission and the fire department with jurisdiction over the facility?   Yes___ No___

23. Are the following hazardous chemicals and substances included on the Hazardous Chemical Inventory forms:
    a. All extremely hazardous substances present in amounts greater than or equal to 500 lb (227 kg, approximately 55 gal) or the threshold planning quantity, whichever is lower?   Yes___ No___
    b. Gasoline (all grades combined) in amounts greater than or equal to 75,000 gal (or approximately 283,900 L) when the gasoline is in tanks entirely underground at a retail gas station that was in compliance during the preceding calendar year with all applicable UST regulations (40 CFR Part 280 or requirements of the state UST program approved by U.S. EPA under 40 CFR Part 281)?   Yes___ No___
    c. Diesel fuel (all grades combined) in amounts greater than or equal to 100,000 gal (or approximately 378,500 L) when the diesel is in tanks entirely underground at a retail gas station that was in compliance during the preceding calendar year with all applicable UST regulations (40 CFR Part 280 or requirements of the state UST program approved by U.S. EPA under 40 CFR Part 281)?   Yes___ No___

(Continued)

TABLE D.2 (CONTINUED)
EPCRA Release/Notification Reporting Checklist

    d. All other hazardous chemicals present at any one time in amounts equal to or greater than 10,000 lb (4,540 kg)?   Yes___ No___

24. Are Tier I or Tier II forms submitted on or before March 1 of the first year after the facility becomes subject to 40 CFR 370.20 through 370.28?   Yes___ No___

    Facilities that manufacture, process or otherwise use a listed toxic chemical in excess of applicable threshold quantities and that have ten or more employees are subject to certain reporting requirements (40 CFR 372.22 through 372.38 and 372.95(b)).

25. Does the facility meet *all* of the following reporting criteria for a calendar year?
    a. The facility has ten or more full-time employees?   Yes___ No___
    b. The facility is in Standard Industrial Classification (SIC) (as in effect on January 1, 1987) major group codes 10 (except 1011, 1081 and 1094), 12 (except 1241) or 20 through 39; industry codes 4911, 4931 or 4939 (limited to facilities that combust coal and/or oil for the purpose of generating power for distribution in commerce); or 4953 (limited to facilities regulated under the RCRA, Subtitle C, 42 USC Section 6921 et seq.), or 5169, or 5171 or 7389 (limited to facilities primarily engaged in solvent recovery services on a contract or fee basis?   Yes___ No___
    c. The facility manufactured (including imported), processed or otherwise used a listed toxic chemical in excess of an applicable threshold quantity of that chemical?   Yes___ No___

26. Is a completed U.S. EPA Form R (US EPA Form 9350-1) submitted annually, for each toxic chemical known by the facility owner or operator to be manufactured (including imported) or otherwise used and exceeding threshold levels in one calendar year to the U.S. EPA and state on or before July 1 of the next year?   Yes___ No___

27. If an alternative reporting threshold are used, was the required certification statement submitted that contains the following information instead of the U.S. EPA Form R – reporting year: Yes___ No___
    a. An indication of whether the chemical identified is being claimed as trade secret
    b. Chemical name and CAS number (if applicable) of the chemical, or the category name
    c. Signature of a senior management official certifying the following: pursuant to 40 CFR 372.27, 'I hereby certify that to the best of my knowledge and belief for the toxic chemical listed in this statement, the annual reportable amount, as defined in 40 CFR 372.27(a), did not exceed 500 lb. for this reporting year and that the chemical was manufactured, or processed, or otherwise used in an amount not exceeding 1 million pounds during this reporting year.'
    d. Date signed
    e. Facility name and address
    f. Mailing address of the facility if different than the above
    g. Toxic chemical release inventory facility identification number if known – name and telephone number of a technical contact
    h. The four-digit SIC codes for the facility or establishments in the facility – latitude and longitude coordinates for the facility
    i. Dun and Bradstreet number of the facility
    j. U.S. EPA identification number(s) (RCRA) I.D. Number(s) of the facility
    k. Facility NPDES permit number(s)
    l. Underground Injection Well Code (UIC) I.D. Number(s) of the facility
    m. Name of the facility's parent company
    n. Parent company's Dun and Bradstreet Number

28. When more than one threshold applies to facility activities, are reports made if any applicable threshold is exceeded?   Yes___ No___

29. If using, producing or manufacturing more than one member of a chemical category listed in 40 CFR 372.65(c), are reports made for exceedances of any applicable threshold for the total volume of all the members of the category involved in the applicable activity and does the report covers all activities at the facility involving members of the category?   Yes___ No___

TABLE D.3
EPCRA Record-Keeping Checklist

Date: _____ Time:_____

Inspector's name: _____ Signature: _____

1. Does the facility manufacture, process or otherwise use a listed toxic chemical in excess of applicable threshold quantities and that have ten or more employees?   Yes___ No___

   If yes to #1, facility is subject to certain record-keeping requirements (40 CFR 372.22(a), 372.22(b), 372.22(c), 372.25(a), 372.25(b), 372.10(a) through 372.10(d) and 372.38).

2. Are the following records kept three years from the date of the submission of U.S. EPA Form R (U.S. EPA Form 9350-1)?

   a. A copy of each Form R report submitted   Yes___ No___

   b. All supporting materials and documentation used by the person to make the compliance determination that the facility or establishments is a covered facility under 40 CFR 372.22 or 372.45   Yes___ No___

3. Is the following documentation supporting the submitted report kept for three years?

   a. Documentation supporting any determination that a claimed allowable exemption under 40 CFR 372.38 applies?   Yes___ No___

   b. Data supporting the determination of whether a reporting threshold applies for each toxic chemical?   Yes___ No___

   c. Documentation supporting the calculations of the quantity of each toxic chemical released to the environment or transferred to an off-site location?   Yes___ No___

   d. Documentation supporting the use indications and quantity on-site reporting for each toxic chemical, including dates of manufacturing, processing or use?   Yes___ No___

   e. Documentation supporting the basis of estimate used in developing any release or off-site transfer estimates for each toxic chemical?   Yes___ No___

   f. Receipts or manifests associated with the transfer of each toxic chemical in waste to off-site locations?   Yes___ No___

   g. Documentation supporting reported waste treatment methods, estimates of treatment efficiencies, ranges of influent concentration to such treatment, the sequential nature of treatment steps, if applicable, and the actual operating data, if applicable, to support the waste treatment efficiency estimate for each toxic chemical?   Yes___ No___

4. Are the following records maintained for three years at the facility to which the report applies or from which supplier notification was provided?

   a. All supporting materials and documentation used to determine if supplier notification is required?   Yes___ No___

   b. All supporting materials and documentation used in developing each required supplier notification and a copy of each notification?   Yes___ No___

5. Has a determination been made that the alternative threshold (see 40 CFR 372.27) may be applied?   Yes___ No___

6. If yes to #5 above, are the following records kept for three years from the date of submission of the required certification statement:

   a. A copy of each certification statement submitted?   Yes___ No___

   b. All supporting materials and documentation used to make the compliance determination that the facility or establishment is eligible to apply the alternative threshold?   Yes___ No___

   c. Documentation supporting the certification statement submitted, including:

   i. Data supporting the determination of whether the alternative threshold applies for each toxic chemical?   Yes___ No___

   ii. Documentation supporting the calculation of annual reportable amount (see 40 CFR 372.37(a)) for each toxic chemical, including documentation supporting the calculations and the calculations of each data element combined for the annual reportable amount?   Yes___ No___

   iii. Receipts or manifests associated with the transfer of each chemical in waste to off-site locations?   Yes___ No___

APPENDIX E
Toolkit for Evaluating SDWA

This appendix provides a general checklist intended to assist compliance managers and personnel performing inspections and assessments of drinking water systems that provide potable drinking water. These tools are provided for instructional purpose and guidance only and are not intended to represent an all-inclusive collection of SWDA regulatory requirements. The checklist provided in this appendix is adapted from the U.S. government agencies. Most states are authorized to implement and enforce SWDA regulations, and many states have state-specific compliance tools available on their websites.

TABLE E.1
SDWA Checklist

Date: Time:
Inspector's name: Signature:
 1. Approximately how many people are served by the facility's drinking water treatment system? (40 CFR 141.2) _____
 2. What is the water treatment system design capacity? (40 CFR 141.2)_____
 3. What chemicals are used for water treatment? (40 CFR 141.74) _____
 4. How much chemical is stored on-site? _____ lbs.
 5. Does this amount exceed SARA TPQs? (40 CFR 355 App. A and App. B)   Yes___ No___
 6. Is the water treatment system approved or permitted by State regulators? (40 CFR 142.4)   Yes___ No___
 7. Who is the State regulating agency for safe drinking water standards?
 8. Do State regulators periodically conduct water quality sampling at your facility? (40 CFR 143.4)   Yes___ No___
 9. Has this facility been found in violation by the regulating agency(s) within the past five years regarding water quality or treatment? (40 CFR 141.32; 141.202– 205)   Yes___ No___
10. If yes or unsure, please explain: _____
_____
_____

11. Does this facility meet all current requirements of the regulating agencies regarding safe drinking water? (40 CFR 141.40; 141.41; 141.42; 141.43)   Yes___ No___
12. If no or unsure, please explain: _____
_____
13. Is the water supply tested regularly?(40 CFR 142.34)   Yes___ No___
14. Is the testing in full compliance with regulatory standards?(40 CFR 142.34)   Yes___ No___
15. Are all water quality samples analyzed by a State Certified lab? (40 CFR 141.28)   Yes___ No___
16. List the analytes (contaminants) that are tested and testing intervals: _____
_____

17. Are detailed records kept regarding system design, operation problems and modifications? (40 CFR 141.91)   Yes___ No___
18. Does the person responding to this checklist feel that this facility is meeting all of its permit requirements regarding SDW standards?   Yes___ No___
    If no, please explain:
19. Does your facility have on-site or manage other facilities that have Class V wells?   Yes___ No___
20. If yes, is there an inventory of these wells that is maintained on-site? (40 CFR 144.26(b)1)   Yes___ No___
21. Has that inventory been submitted to the appropriate State or EPA managing authority? (40 CFR 144.83)   Yes___ No___
22. Date of inventory submitted?
23. Do any of the wells managed by this facility require a specific permit? (40 CFR 144.84)   Yes___ No___

Class V Large-Capacity Cesspools and Motor Vehicle Waste Disposal Wells
24. Do any of these wells qualify as large-capacity cesspools or motor vehicle waste disposal wells? (40 CFR 144.3 and 40 CFR 144.85)   Yes___ No___
25. Are any of these wells located within groundwater protection areas or other sensitive groundwater areas? (40 CFR 144.87)   Yes___ No___
    (If yes, there are additional reporting and permitting requirements for motor vehicle waste disposal wells.)
26. Do (or did) any of these wells require closure? (40 CFR §144.89)   Yes___ No___
    Additional comments: _____
_____
_____

APPENDIX F
Toolkit for Evaluating SPCC

This appendix provides a series of checklists and evaluation tools intended to assist compliance managers and personnel performing inspections and assessments of some of the more common operations involving planning, implementing, inspecting and record-keeping governed by SPCC. These tools are provided for instructional purpose and guidance only and are not intended to represent an all-inclusive collection of SPCC regulatory requirements. The tools provided in this appendix are adapted from EPA and organized within this appendix as follows.

SPCC:

1. Tier I Facility
2. Tier II Onshore Facility

TABLE F-1 Tier I Facility Checklist

SPCC GENERAL APPLICABILITY—40 CFR 112.1	
IS THE FACILITY REGULATED UNDER 40 CFR part 112?	
The completely buried oil storage capacity is over 42,000 U.S. gallons, **OR** the aggregate aboveground oil storage capacity is over 1,320 U.S. gallons **AND**	☐ Yes ☐ No ☐ Yes ☐ No
The facility is a non-transportation-related facility engaged in drilling, producing, gathering, storing, processing, refining, transferring, distributing, using, or consuming oil and oil products, which due to its location could reasonably be expected to discharge oil into or upon the navigable waters of the United States	

AFFECTED WATERWAY(S):	DISTANCE:
FLOW PATH TO WATERWAY:	

Note: The following storage capacity is not considered in determining applicability of SPCC requirements:

- *Equipment subject to the authority of the U.S. Department of Transportation, U.S. Department of the Interior, or Minerals Management Service, as defined in Memoranda of Understanding dated November 24, 1971, and November 8, 1993; Tank trucks that return to an otherwise regulated facility that contain only residual amounts of oil (EPA Policy letter)*
- *Completely buried tanks subject to all the technical requirements of 40 CFR part 280 or a state program approved under 40 CFR part 281;*
- *Underground oil storage tanks deferred under 40 CFR part 280 that supply emergency diesel generators at a nuclear power generation facility licensed by the Nuclear Regulatory Commission (NRC) and subject to any NRC provision regarding design and quality criteria, including but not limited to CFR part 50;*
- *Any facility or part thereof used exclusively for wastewater treatment (production, recovery or recycling of oil is not considered wastewater treatment): (This does not include other oil containers located at a wastewater treatment facility, such as generator tanks or transformers)*

- *Containers smaller than 55 U.S. gallons;*
- *Permanently closed containers (as defined in §112.2);*
- *Motive power containers (as defined in §112.2);*
- *Hot-mix asphalt or any hot-mix asphalt containers;*
- *Heating oil containers used solely at a single-family residence;*
- *Pesticide application equipment and related mix containers;*
- *Any milk and milk product container and associated piping and appurtenances; and*
- *Intra-facility gathering lines subject to the regulatory requirements of 49 CFR part 192 or 195.*

Does the facility have an SPCC Plan?	☐ Yes ☐ No

SPCC TIER I QUALIFIED FACILITY APPLICABILITY—40 CFR 112.3(g)(1),(2)	
The aggregate aboveground oil storage capacity is 10,000 U.S. gallons or less **AND**	☐ Yes ☐ No
The capacity of each individual aboveground oil storage container is 5,000 U.S. gallons or less **AND**	☐ Yes ☐ No
In the three years prior to the SPCC Plan self-certification date, or since becoming subject to the rule (if the facility has been in operation for less than three years), the facility has **NOT** had:	
• A single discharge as described in §112.1(b) exceeding 1,000 U.S. gallons, **OR**	☐ Yes ☐ No
• Two discharges as described in §112.1(b) each exceeding 42 U.S. gallons within any twelve-month period[1]	☐ Yes ☐ No
IF **YES** TO ALL OF THE ABOVE, THEN THE FACILITY IS CONSIDERED A TIER I QUALIFIED FACILITY.[2]	

Comments:

---

[1] Oil discharges that result from natural disasters, acts of war, or terrorism are not included in this determination. The gallon amount(s) specified (either 1,000 or 42) refers to the amount of oil that actually reaches navigable waters or adjoining shorelines not the total amount of oil spilled. The entire volume of the discharge is oil for this determination.
[2] An owner/operator who self-certifies a Tier I SPCC Plan may not include any environmentally equivalent alternatives or secondary containment impracticability determinations in the SPCC Plan

REQUIREMENTS FOR PREPARATION AND IMPLEMENTATION OF A SPCC PLAN—40 CFR 112.3

Date facility began operations:

Date of initial SPCC Plan preparation:	Current Plan version (date/number):

112.3(a)	**For facilities (except farms), including mobile or portable facilities:** • In operation on or prior to November 10, 2011: Plan prepared and/or amended and fully implemented by **November 10, 2011**	☐ Yes ☐ No ☐ NA
	• Facilities beginning operation after November 10, 2011:	
	○ *Oil production facilities* - Plan prepared and fully implemented within six months after beginning operations; or	☐ Yes ☐ No ☐ NA
	○ *All other facilities* - Plan prepared and fully implemented before operations begin	☐ Yes ☐ No ☐ NA
	For farms (as defined in §112.2): • In operation on or prior to August 16, 2002: Plan maintained, amended and implemented by **May 10, 2013**	☐ Yes ☐ No ☐ NA
	• Beginning operations after August 16, 2002 through May 10, 2013: Plan prepared and fully implemented by **May 10, 2013**	☐ Yes ☐ No ☐ NA
	• Beginning operations after May 10, 2013: Plan prepared and fully implemented before beginning operations	☐ Yes ☐ No ☐ NA
112.3(e)(1)	Plan is available onsite if attended at least 4 hours per day. If facility is unattended, Plan is available at the nearest field office. *(Please note nearest field office contact information in comments section below.)*	☐ Yes ☐ No ☐ NA

Comments:

AMENDMENT OF SPCC PLAN BY REGIONAL ADMINISTRATOR (RA)—40 CFR 112.4

112.4(a),(c)	Has the facility discharged more than 1,000 U.S. gallons of oil in a single reportable discharge or more than 42 U.S. gallons in each of two reportable discharges in any 12-month period?[3]	☐ Yes ☐ No
If YES	• Was information submitted to the RA as required in §112.4(a)?[4]	☐ Yes ☐ No ☐ NA
	• Was information submitted to the appropriate agency or agencies in charge of oil pollution control activities in the State in which the facility is located§112.4(c)	☐ Yes ☐ No ☐ NA
	• Date(s) and volume(s) of reportable discharges(s) under this section: _____	
	• Were the discharges reported to the NRC[5]?	☐ Yes ☐ No
112.4(d),(e)	Have changes required by the RA been implemented in the Plan and/or facility?	☐ Yes ☐ No ☐ NA

Comments:

---

[3] A reportable discharge is a discharge as described in §112.1(b)(see 40 CFR part 110). The gallon amount(s) specified (either 1,000 or 42) refers to the amount of oil that actually reaches navigable waters or adjoining shorelines not the total amount of oil spilled. The entire volume of the discharge is oil for this determination

[4] Triggering this threshold may disqualify the facility from meeting the Qualified Facility criteria if it occurred in the three years prior to self-certification

[5] Inspector Note-Confirm any spills identified above were reported to NRC

AMENDMENT OF SPCC PLAN BY THE OWNER OR OPERATOR—40 CFR 112.5

112.5(a)	Has there been a change at the facility that materially affects the potential for a discharge described in §112.1(b)?	☐ Yes ☐ No
If YES	• Was the Plan amended within six months of the change?	☐ Yes ☐ No
	• Were amendments implemented within six months of any Plan amendment?	☐ Yes ☐ No
112.5(b)	Review and evaluation of the Plan completed at least once every 5 years?	☐ Yes ☐ No ☐ NA
	Following Plan review, was Plan amended within six months to include more effective prevention and control technology that has been field-proven to significantly reduce the likelihood of a discharge described in §112.1(b)?	☐ Yes ☐ No ☐ NA
	Amendments implemented within six months of any Plan amendment?	☐ Yes ☐ No ☐ NA
	Five year Plan review and evaluation documented?	☐ Yes ☐ No ☐ NA
112.5(c)	Professional Engineer certification of any technical Plan amendments in accordance with all applicable requirements of §112.3(d) *[Except for self-certified Plans]*	☐ Yes ☐ No ☐ NA

Name:	License No.:	State:	Date of certification:

Reason for amendment:

TIER I QUALIFIED FACILITY PLAN REQUIREMENTS —40 CFR 112.6(a)

112.6(a)(1)	**Plan Certification:** Plan prepared to comply with the requirements of §112.6(a)(3) using the Appendix G template	☐ Yes ☐ No ☐ NA
(i)	He or she is familiar with the requirements of 40 CFR part 112	☐ Yes ☐ No ☐ NA
(ii)	He or she has visited and examined the facility[6]	☐ Yes ☐ No ☐ NA
(iii)	The Plan has been prepared in accordance with accepted and sound industry practices and standards	☐ Yes ☐ No ☐ NA
(iv)	Procedures for required inspections and testing have been established	☐ Yes ☐ No ☐ NA
(v)	He or she will fully implement the Plan	☐ Yes ☐ No ☐ NA
(vi)	The facility meets the qualification criteria in §112.3(g)(1)	☐ Yes ☐ No ☐ NA
(vii)	The Plan does not deviate from any requirements as allowed by §§112.7(a)(2) and 112.7(d), or include measures pursuant to §112.9(c)(6) for produced water containers and any associated piping	☐ Yes ☐ No ☐ NA
(viii)	The Plan and individual(s) responsible for implementing the Plan have the full approval of management and the facility owner or operator has committed the necessary resources to fully implement the Plan.	☐ Yes ☐ No ☐ NA
112.6(a)(2)	**Technical Amendments:** The owner/operator self-certified the Plan's technical amendments for a change in facility design, construction, operation, or maintenance that affected potential for a §112.1(b) discharge	☐ Yes ☐ No ☐ NA
If YES	• Certification of technical amendments is in accordance with the self-certification provisions of §112.6(a)(1).	☐ Yes ☐ No ☐ NA
	An individual oil storage container capacity exceeds 5,000 U.S. gallons or the aggregate aboveground oil storage capacity increased to more than 10,000 U.S. gallons as a result of the change	☐ Yes ☐ No ☐ NA
If YES	*The facility no longer meets the Tier I qualifying criteria in §112.3(g)(1) because an individual oil storage container capacity exceeds 5,000 U.S. gallons or the facility aboveground storage capacity exceeds 10,000 U.S. gallons*	
	The following has been or will be completed within six months following the amendment:	
(i)	• Plan prepared and implemented in accordance with the requirements for a Tier II Qualified Facility (§112.6(b)) if the facility meets the eligibility criteria **OR**	☐ Yes ☐ No ☐ NA
(ii)	• Plan prepared and implemented in accordance with the general Plan requirements in §112.7 and applicable requirements in subparts B and C and certified by a PE as required under §112.3(d)	☐ Yes ☐ No ☐ NA

---

[6] Note that only the person certifying the Plan can make the site visit

		PLAN	
112.6(a)(3)(i)	Plan includes a prediction of the direction and total quantity of oil which could be discharged from the facility as a result of each type of major equipment failure if there is a reasonable potential for equipment failure (such as loading or unloading equipment, tank overflow, rupture, or leakage, or any other equipment known to be a source of discharge)	☐ Yes ☐ No ☐ NA	
(ii)	Bulk storage container installations (except mobile refuelers and other non-transportation-related tank trucks), including mobile or portable oil storage containers, are constructed to provide secondary containment for the entire capacity of the largest single container plus additional capacity to contain precipitation, and	☐ Yes ☐ No ☐ NA	
	Mobile or portable oil storage containers positioned or located to prevent a §112.1(b) discharge	☐ Yes ☐ No ☐ NA	
(iii)	Plan describes a system or documented procedure to prevent overfills for each container and is regularly tested to ensure proper operation or efficacy	☐ Yes ☐ No ☐ NA	

Comments:

GENERAL SPCC REQUIREMENTS—40 CFR 112.7		PLAN	FIELD
Management approval at a level of authority to commit the necessary resources to fully implement the Plan[7]		☐ Yes ☐ No	
Plan follows sequence of the rule or is an equivalent Plan meeting all applicable rule requirements and includes a cross-reference of provisions		☐ Yes ☐ No ☐ NA	
If Plan calls for facilities, procedures, methods, or equipment not yet fully operational, details of their installation and start-up are discussed *(Note: Relevant for inspection evaluation and testing baselines.)*		☐ Yes ☐ No ☐ NA	
112.7(a)(3)	Plan addresses each of the following:		
(i)	For each fixed container, type of oil and storage capacity (see Attachment C of this checklist). For mobile or portable containers, type of oil and storage capacity for each container or an estimate of the potential number of mobile or portable containers, the types of oil, and anticipated storage capacities	☐ Yes ☐ No	☐ Yes ☐ No
(iv)	Countermeasures for discharge discovery, response, and cleanup (both facility's and contractor's resources)	☐ Yes ☐ No	☐ Yes ☐ No
(vi)	Contact list and phone numbers for the facility response coordinator, National Response Center, cleanup contractors with an agreement for response, and all Federal, State, and local agencies who must be contacted in the case of a discharge as described in §112.1(b)	☐ Yes ☐ No	
112.7(a)(4)	Plan includes information and procedures that enable a person reporting an oil discharge as described in §112.1(b) to relate information on the: • Exact address or location and phone number of the facility; • Date and time of the discharge; • Type of material discharged; • Estimates of the total quantity discharged; • Estimates of the quantity discharged as described in §112.1(b); • Source of the discharge; • A description of all affected media; • Cause of the discharge; • Damages or injuries caused by the discharge; • Actions being used to stop, remove, and mitigate the effects of the discharge; • Whether an evacuation may be needed; • Names of individuals and/or organizations who have also been contacted	☐ Yes ☐ No ☐ NA	
112.7(a)(5)	Plan organized so that portions describing procedures to be used when a discharge occurs will be readily usable in an emergency	☐ Yes ☐ No ☐ NA	

Comments:

		PLAN	FIELD

---

[7] May be part of the Plan or demonstrated elsewhere.

		PLAN	FIELD
112.7(c)	Appropriate containment and/or diversionary structures or equipment are provided to prevent a discharge as described in §112.1(b), **except as provided in §112.7(k) of this section for certain qualified operational equipment and §112.9(d)(3) for certain flowlines and intra-facility gathering lines at an oil production facility.** The entire containment system, including walls and floors, are capable of containing oil and are constructed to prevent escape of a discharge from the containment system before cleanup occurs. The method, design, and capacity for secondary containment address the typical failure mode and the most likely quantity of oil that would be discharged. See Attachment C of this checklist. **For onshore facilities,** one of the following or its equivalent: • Dikes, berms, or retaining walls sufficiently impervious to contain oil, • Weirs, booms or other barriers, • Curbing or drip pans, • Spill diversion ponds, • Sumps and collection systems, • Retention ponds, or • Culverting, gutters or other drainage systems, • Sorbent materials		
	Identify which of the following are present at the facility and if appropriate containment and/or diversionary structures or equipment are provided as described above:		
	☐ Bulk storage containers	☐ Yes ☐ No ☐ NA	☐ Yes ☐ No ☐ NA
	☐ Mobile/portable containers	☐ Yes ☐ No ☐ NA	☐ Yes ☐ No ☐ NA
	☐ Oil-filled operational equipment (as defined in 112.2)	☐ Yes ☐ No ☐ NA	☐ Yes ☐ No ☐ NA
	☐ Other oil-filled equipment (i.e., manufacturing equipment)	☐ Yes ☐ No ☐ NA	☐ Yes ☐ No ☐ NA
	☐ Piping and related appurtenances	☐ Yes ☐ No ☐ NA	☐ Yes ☐ No ☐ NA
	☐ Mobile refuelers or non-transportation-related tank cars	☐ Yes ☐ No ☐ NA	☐ Yes ☐ No ☐ NA
	☐ Transfer areas, equipment and activities	☐ Yes ☐ No ☐ NA	☐ Yes ☐ No ☐ NA
	☐ Identify any other equipment or activities that are not listed above: _____	☐ Yes ☐ No ☐ NA	☐ Yes ☐ No ☐ NA
112.7(e)	Inspections and tests conducted in accordance with written procedures	☐ Yes ☐ No	☐ Yes ☐ No
	Record of inspections or tests signed by supervisor or inspector	☐ Yes ☐ No	☐ Yes ☐ No
	Kept with Plan for at least 3 years (see Attachment D of this checklist)[8]	☐ Yes ☐ No	☐ Yes ☐ No
112.7(f)	Personnel, training, and oil discharge prevention procedures		
(1)	Training of oil-handling personnel in operation and maintenance of equipment to prevent discharges; discharge procedure protocols; applicable pollution control laws, rules, and regulations; general facility operations; and contents of SPCC Plan	☐ Yes ☐ No ☐ NA	☐ Yes ☐ No ☐ NA
(2)	Person designated as accountable for discharge prevention at the facility and reports to facility management	☐ Yes ☐ No ☐ NA	☐ Yes ☐ No ☐ NA
(3)	Discharge prevention briefings conducted at least once a year for oil handling personnel to assure adequate understanding of the Plan. Briefings highlight and describe known discharges as described in §112.1(b) or failures, malfunctioning components, and any recently developed precautionary measures	☐ Yes ☐ No ☐ NA	☐ Yes ☐ No ☐ NA
Comments:			
		PLAN	FIELD

[a] Records of inspections and tests kept under usual and customary business practices will suffice

112.7(g)	Plan describes how to: • Secure and control access to the oil handling, processing and storage areas; • Secure master flow and drain valves; • Prevent unauthorized access to starter controls on oil pumps; • Secure out-of-service and loading/unloading connections of oil pipelines; and • Address the appropriateness of security lighting to both prevent acts of vandalism and assist in the discovery of oil discharges	☐ Yes ☐ No ☐ NA ☐ Yes ☐ No ☐ NA *For Oil Production Facilities:* *Select NA*	
112.7(k)	Qualified oil-filled operational equipment is present at the facility[9] *Oil-filled operational equipment means* equipment that includes an oil storage container (or multiple containers) in which the oil is present solely to support the function of the apparatus or the device. Oil-filled operational equipment is not considered a bulk storage container, and does not include oil-filled manufacturing equipment (flow-through process). Examples of oil-filled operational equipment include, but are not limited to, hydraulic systems, lubricating systems (e.g. , those for pumps, compressors and other rotating equipment, including pumpjack lubrication systems), gear boxes, machining coolant systems, heat transfer systems, transformers, circuit breakers, electrical switches, and other systems containing oil solely to enable the operation of the device.	☐ Yes ☐ No	
If YES	Check which apply: Secondary Containment provided in accordance with 112.7(c) ☐ Alternative measure described below (confirm eligibility) ☐		
112.7(k)	Qualified Oil-Filled Operational Equipment • Has a single reportable discharge as described in §112.1(b) from any oil-filled operational equipment exceeding 1,000 U.S. gallons occurred within the three years prior to Plan certification date?	☐ Yes ☐ No ☐ NA	
	• Have two reportable discharges as described in §112.1(b) from any oil-filled operational equipment each exceeding 42 U.S. gallons occurred within any 12-month period within the three years prior to Plan certification date?[10]	☐ Yes ☐ No ☐ NA	
	If **YES** for either, secondary containment in accordance with §112.7(c) is required		
	• Facility procedure for inspections or monitoring program to detect equipment failure and/or a discharge is established and documented ***Does not apply if the facility has submitted a FRP under §112.20:*** • Contingency plan following 40 CFR part 109 (see Attachment E of this checklist) is provided in Plan **AND** • Written commitment of manpower, equipment, and materials required to expeditiously control and remove any quantity of oil discharged that may be harmful is provided in Plan	☐ Yes ☐ No ☐ NA ☐ Yes ☐ No ☐ NA ☐ Yes ☐ No ☐ NA	☐ Yes ☐ No ☐ NA
Comments:			
Note- Complete, as applicable, either Attachment A or B which include additional requirements based on the type of facility.			

[9] This provision does not apply to oil-filled manufacturing equipment (flow-through process)

[10] Oil discharges that result from natural disasters, acts of war, or terrorism are not included in this determination. The gallon amount(s) specified (either 1,000 or 42) refers to the amount of oil that actually reaches navigable waters or adjoining shorelines not the total amount of oil spilled. The entire volume of the discharge is oil for this determination.

ATTACHMENT A ONSHORE FACILITIES (EXCLUDING PRODUCTION) 40 CFR 112.8/112.12 ☐ NA	PLAN	FIELD
112.8(b)/ 112.12(b) Facility Drainage		
Diked Areas (1) Drainage from diked storage areas is: • Restrained by valves, except where facility systems are designed to control such discharge, **OR** • Manually activated pumps or ejectors are used and the condition of the accumulation is inspected prior to draining dike to ensure no oil will be discharged	☐ Yes ☐ No ☐ NA	☐ Yes ☐ No ☐ NA

Comments:

112.8(c)/112.12(c) Bulk Storage Containers   ☐ NA

Bulk storage container means any container used to store oil. These containers are used for purposes including, but not limited to, the storage of oil prior to use, while being used, or prior to further distribution in commerce. Oil-filled electrical, operating, or manufacturing equipment is not a bulk storage container.

If bulk storage containers are not present, mark this section Not Applicable (NA). If present, complete this section and Attachment C of this checklist.

	PLAN	FIELD
(1) Containers materials and construction are compatible with material stored and conditions of storage such as pressure and temperature	☐ Yes ☐ No ☐ NA	☐ Yes ☐ No ☐ NA
(3) Is there drainage of uncontaminated rainwater from diked areas into a storm drain or open watercourse?	☐ Yes ☐ No ☐ NA	☐ Yes ☐ No ☐ NA
If **YES** Bypass valve normally sealed closed	☐ Yes ☐ No ☐ NA	☐ Yes ☐ No ☐ NA
• Retained rainwater is inspected to ensure that its presence will not cause a discharge as described in §112.1(b)	☐ Yes ☐ No ☐ NA	☐ Yes ☐ No ☐ NA
• Bypass valve opened and resealed under responsible supervision	☐ Yes ☐ No ☐ NA	☐ Yes ☐ No ☐ NA
• Adequate records of drainage are kept; for example, records required under permits issued in accordance with 40 CFR §§122.41(j)(2) and (m)(3)	☐ Yes ☐ No ☐ NA	☐ Yes ☐ No ☐ NA
(4) For completely buried metallic tanks installed on or after January 10, 1974 (if not exempt from SPCC regulation because subject to all of the technical requirements of 40 CFR part 280 or 281):		
• Provide corrosion protection with coatings or cathodic protection compatible with local soil conditions	☐ Yes ☐ No ☐ NA	☐ Yes ☐ No ☐ NA
• Regular leak testing conducted	☐ Yes ☐ No ☐ NA	☐ Yes ☐ No ☐ NA
(5) The buried section of partially buried or bunkered metallic tanks protected from corrosion with coatings or cathodic protection compatible with local soil conditions	☐ Yes ☐ No ☐ NA	☐ Yes ☐ No ☐ NA

Comments:

ATTACHMENT A		PLAN	FIELD
(6)	Test or inspect each aboveground container for integrity on a regular schedule and whenever you make material repairs. Techniques include, but are not limited to: visual inspection, hydrostatic testing, radiographic testing, ultrasonic testing, acoustic emissions testing, or other system of non-destructive testing	☐ Yes ☐ No ☐ NA	☐ Yes ☐ No ☐ NA
	Appropriate qualifications for personnel performing tests and inspections are identified in the Plan and have been assessed in accordance with industry standards	☐ Yes ☐ No ☐ NA	☐ Yes ☐ No ☐ NA
	• The frequency and type of testing and inspections are documented, are in accordance with industry standards and take into account the container size, configuration and design	☐ Yes ☐ No ☐ NA	☐ Yes ☐ No ☐ NA
	• Comparison records of aboveground container integrity testing are maintained	☐ Yes ☐ No ☐ NA	☐ Yes ☐ No ☐ NA
	• Container supports and foundations regularly inspected	☐ Yes ☐ No ☐ NA	☐ Yes ☐ No ☐ NA
	• Outside of containers frequently inspected for signs of deterioration, discharges, or accumulation of oil inside diked areas	☐ Yes ☐ No ☐ NA	☐ Yes ☐ No ☐ NA
	• Records of all inspections and tests maintained[11]	☐ Yes ☐ No ☐ NA	☐ Yes ☐ No ☐ NA

Integrity Testing Standard identified in the Plan:

112.12 (c)(6)(ii) *(Applies to AFVO Facilities only)*	Conduct formal visual inspection on a regular schedule for bulk storage containers that meet all of the following conditions: • Subject to 21 CFR part 110; • Have no external insulation; and • Elevated; • Shop-fabricated. • Constructed of austenitic stainless steel;	☐ Yes ☐ No ☐ NA	☐ Yes ☐ No ☐ NA
	In addition, you must frequently inspect the outside of the container for signs of deterioration, discharges, or accumulation of oil inside diked areas.	☐ Yes ☐ No ☐ NA	☐ Yes ☐ No ☐ NA
	You must determine and document in the Plan the appropriate qualifications for personnel performing tests and inspections.[11]	☐ Yes ☐ No ☐ NA	☐ Yes ☐ No ☐ NA
(10)	Visible discharges which result in a loss of oil from the container, including but not limited to seams, gaskets, piping, pumps, valves, rivets, and bolts are promptly corrected and oil in diked areas is promptly removed	☐ Yes ☐ No ☐ NA	☐ Yes ☐ No ☐ NA

112.8(d)/112.12(d)Facility transfer operations, pumping, and facility process

(4)	Aboveground valves, piping, and appurtenances such as flange joints, expansion joints, valve glands and bodies, catch pans, pipeline supports, locking of valves, and metal surfaces are inspected regularly to assess their general condition	☐ Yes ☐ No ☐ NA	☐ Yes ☐ No ☐ NA
	Integrity and leak testing conducted on buried piping at time of installation, modification, construction, relocation, or replacement	☐ Yes ☐ No ☐ NA	☐ Yes ☐ No ☐ NA

Comments:

[11] Records of inspections and tests kept under usual and customary business practices will suffice

ATTACHMENT B ☐ NA	PLAN	FIELD
ONSHORE OIL PRODUCTION FACILITIES—40 CFR 112.9		

(Drilling and workover facilities are excluded from the requirements of §112.9)

Production facility means all structures (including but not limited to wells, platforms, or storage facilities), piping (including but not limited to flowlines or intra-facility gathering lines), or equipment (including but not limited to workover equipment, separation equipment, or auxiliary non-transportation-related equipment) used in the production, extraction, recovery, lifting, stabilization, separation or treating of oil (including condensate), or associated storage or measurement, and is located in an oil or gas field, at a facility. This definition governs whether such structures, piping, or equipment are subject to a specific section of this part.

112.9(b) Oil Production Facility Drainage

		PLAN	FIELD
(1)	At tank batteries, separation and treating areas where there is a reasonable possibility of a discharge as described in §112.1(b), drains for dikes or equivalent measures are closed and sealed except when draining uncontaminated rainwater. Accumulated oil on the rainwater is removed and then returned to storage or disposed of in accordance with legally approved methods	☐ Yes ☐ No ☐ NA	☐ Yes ☐ No ☐ NA
	Prior to drainage, diked area inspected and action taken as provided below:		
	• 112.8(c)(3)(ii) - Retained rainwater is inspected to ensure that its presence will not cause a discharge as described in §112.1(b)	☐ Yes ☐ No ☐ NA	☐ Yes ☐ No ☐ NA
	• 112.8(c)(3)(iii) - Bypass valve opened and resealed under responsible supervision	☐ Yes ☐ No ☐ NA	☐ Yes ☐ No ☐ NA
	• 112.8(c)(3)(iv) - Adequate records of drainage are kept; for example, records required under permits issued in accordance with §122.41(j)(2) and (m)(3)	☐ Yes ☐ No ☐ NA	☐ Yes ☐ No ☐ NA
(2)	Field drainage systems (e.g., drainage ditches or road ditches) and oil traps, sumps, or skimmers inspected at regularly scheduled intervals for oil, and accumulations of oil promptly removed	☐ Yes ☐ No ☐ NA	☐ Yes ☐ No ☐ NA

112.9(c) Oil Production Facility Bulk Storage Containers

Bulk storage container means any container used to store oil. These containers are used for purposes including, but not limited to, the storage of oil prior to use, while being used, or prior to further distribution in commerce. Oil-filled electrical, operating, or manufacturing equipment is not a bulk storage container.

		PLAN	FIELD
(1)	Containers materials and construction are compatible with material stored and conditions of storage such as pressure and temperature	☐ Yes ☐ No ☐ NA	☐ Yes ☐ No ☐ NA
(2)	Except as allowed for flow-through process vessels in §112.9(c)(5) and produced water containers in §112.9(c)(6), secondary containment provided for all tank battery, separation and treating facilities sized to hold the capacity of largest single container and sufficient freeboard for precipitation.	☐ Yes ☐ No ☐ NA	☐ Yes ☐ No ☐ NA
	Drainage from undiked area safely confined in a catchment basin or holding pond.	☐ Yes ☐ No ☐ NA	☐ Yes ☐ No ☐ NA
(3)	Except as allowed for flow-through process vessels in §112.9(c)(5) and produced water containers in §112.9(c)(6), periodically and upon a regular schedule, visually inspect containers for deterioration and maintenance needs, including foundation and supports of each container on or above the surface of the ground	☐ Yes ☐ No ☐ NA	☐ Yes ☐ No ☐ NA
(4)	New and old tank batteries engineered/updated in accordance with good engineering practices to prevent discharges including at least one of the following: • Adequate container capacity to prevent overfill if a pumper/gauger is delayed in making regularly scheduled rounds; • Overflow equalizing lines between containers so that a full container can overflow to an adjacent container; • Adequate vacuum protection to prevent container collapse; or • High level sensors to generate and transmit an alarm to the computer where the facility is subject to a computer production control system	☐ Yes ☐ No ☐ NA	☐ Yes ☐ No ☐ NA

Comments:

ATTACHMENT B		PLAN	FIELD
(5)	**Flow-through Process Vessels.** Alternate requirements in lieu of sized secondary containment required in (c)(2) and requirements in (c)(3) above for facilities with flow-through process vessels:		
(i)	Flow-through process vessels and associated components (e.g. dump valves) are periodically and on a regular schedule visually inspected and/or tested for leaks, corrosion, or other conditions that could lead to a discharge as described in §112.1(b)	☐ Yes ☐ No ☐ NA	☐ Yes ☐ No ☐ NA
(ii)	Corrective actions or repairs have been made to flow-through process vessels and any associated components as indicated by regularly scheduled visual inspections, tests, or evidence of an oil discharge	☐ Yes ☐ No ☐ NA	☐ Yes ☐ No ☐ NA
(iii)	Oil removed or other actions initiated to promptly stabilize and remediate any accumulation of oil discharges associated with the produced water container	☐ Yes ☐ No ☐ NA	☐ Yes ☐ No ☐ NA
(iv)	All flow-through process vessels comply with §§112.9(c)(2) and (c)(3) within six months of any flow-through process vessel discharge of more than 1,000 U.S. gallons of oil in a single discharge as described in §112.1(b) or discharges of more than 42 U.S. gallons of oil in each of two discharges as described in §112.1(b) within any twelve month period.[12]	☐ Yes ☐ No ☐ NA	☐ Yes ☐ No ☐ NA
112.9(d) Facility transfer operations, pumping, and facility process			
(1)	All aboveground valves and piping associated with transfer operations are inspected periodically and upon a regular schedule to determine their general condition. Include the general condition of flange joints, valve glands and bodies, drip pans, pipe supports, pumping well polish rod stuffing boxes, bleeder and gauge valves, and other such items	☐ Yes ☐ No ☐ NA	☐ Yes ☐ No ☐ NA
(3)	If flowlines and intra-facility gathering lines are not provided with secondary containment in accordance with §112.7(c) and the facility is not required to submit an FRP under §112.20, then the SPCC Plan includes:		
(i)	• An oil spill contingency plan following the provisions of 40 CFR part 109[13]	☐ Yes ☐ No ☐ NA	☐ Yes ☐ No ☐ NA
(ii)	• A written commitment of manpower, equipment, and materials required to expeditiously control and remove any quantity of oil discharged that might be harmful	☐ Yes ☐ No ☐ NA	☐ Yes ☐ No ☐ NA
Comments:			

[12] Oil discharges that result from natural disasters, acts of war, or terrorism are not included in this determination. The gallon amount(s) specified (either 1,000 or 42) refers to the amount of oil that actually reaches navigable waters or adjoining shorelines not the total amount of oil spilled. The entire volume of the discharge is oil for this determination.
[13] Note that the implementation of a 40 CFR part 109 plan does not require a PE impracticability determination for this specific requirement

ATTACHMENT B		PLAN	FIELD
(4)	A flowline/intra-facility gathering line maintenance program to prevent discharges is prepared and implemented and includes the following procedures:		
(i)	Flowlines and intra-facility gathering lines and associated valves and equipment are compatible with the type of production fluids, their potential corrosivity, volume, and pressure, and other conditions expected in the operational environment	☐ Yes ☐ No ☐ NA	☐ Yes ☐ No ☐ NA
(ii)	Flowlines and intra-facility gathering lines and associated appurtenances are visually inspected and/or tested on a periodic and regular schedule for leaks, oil discharges, corrosion, or other conditions that could lead to a discharge as described in §112.1(b).	☐ Yes ☐ No ☐ NA	☐ Yes ☐ No ☐ NA
	If flowlines and intra-facility gathering lines are not provided with secondary containment in accordance with §112.7(c), the frequency and type of testing allows for the implementation of a contingency plan as described under 40 CFR 109 or an FRP submitted under §112.20	☐ Yes ☐ No ☐ NA	☐ Yes ☐ No ☐ NA
(iii)	Repairs or other corrective actions are made to any flowlines and intra-facility gathering lines and associated appurtenances as indicated by regularly scheduled visual inspections, tests, or evidence of a discharge	☐ Yes ☐ No ☐ NA	☐ Yes ☐ No ☐ NA
(iv)	Oil removed or other actions initiated to promptly stabilize and remediate any accumulation of oil discharges associated with the produced water containers	☐ Yes ☐ No ☐ NA	☐ Yes ☐ No ☐ NA

ATTACHMENT B ☐ NA ONSHORE OIL DRILLING AND WORKOVER FACILITIES—40 CFR 112.10		PLAN	FIELD
112.10(b)	Mobile drilling or workover equipment is positioned or located to prevent a discharge as described in §112.1(b)	☐ Yes ☐ No ☐ NA	☐ Yes ☐ No ☐ NA
112.10(c)	Catchment basins or diversion structures are provided to intercept and contain discharges of fuel, crude oil, or oily drilling fluids	☐ Yes ☐ No ☐ NA	☐ Yes ☐ No ☐ NA
112.10(d)	Blowout prevention (BOP) assembly and well control system installed before drilling below any casing string or during workover operations	☐ Yes ☐ No ☐ NA	☐ Yes ☐ No ☐ NA
	BOP assembly and well control system is capable of controlling any well-head pressure that may be encountered while on the well	☐ Yes ☐ No ☐ NA	☐ Yes ☐ No ☐ NA

ATTACHMENT C: SPCC FIELD INSPECTION AND PLAN REVIEW TABLE
Documentation of Field Observations for Containers and Associated Requirements

Inspectors should use this table to document observations of containers as needed.

Containers and Piping

Check containers for leaks, specifically looking for: drip marks, discoloration of tanks, puddles containing spilled or leaked material, corrosion, cracks, and localized dead vegetation, and standards/specifications of construction.

Check aboveground container foundation for: cracks, discoloration, and puddles containing spilled or leaked material, settling, gaps between container and foundation, and damage caused by vegetation roots.

Check all piping for: droplets of stored material, discoloration, corrosion, bowing of pipe between supports, evidence of stored material seepage from valves or seals, evidence of leaks, and localized dead vegetation. For all aboveground piping, include the general condition of flange joints, valve glands and bodies, drip pans, pipe supports, bleeder and gauge valves, and other such items (Document in comments section of §112.8(d) or 112.12(d).)

Secondary Containment (Active and Passive)

Check secondary containment for: containment system (including walls and floor) ability to contain oil such that oil will not escape the containment system before cleanup occurs, proper sizing, cracks, discoloration, presence of spilled or leaked material (standing liquid), erosion, corrosion, penetrations in the containment system, and valve conditions.

Check dike or berm systems for: level of precipitation in dike/available capacity, operational status of drainage valves (closed), dike or berm impermeability, debris, erosion, impermeability of the earthen floor/walls of diked area, and location/status of pipes, inlets, drainage around and beneath containers, presence of oil discharges within diked areas.

Check drainage systems for: an accumulation of oil that may have resulted from any small discharge, including field drainage systems (such as drainage ditches or road ditches), and oil traps, sumps, or skimmers. Ensure any accumulations of oil have been promptly removed.

Check retention and drainage ponds for: erosion, available capacity, presence of spilled or leaked material, debris, and stressed vegetation.

Check active measures (countermeasures) for: amount indicated in plan is available and appropriate; deployment procedures are realistic; material is located so that they are readily available; efficacy of discharge detection; availability of personnel and training, appropriateness of measures to prevent a discharge as described in §112.1(b). *Note that appropriate evaluation and consideration must be given to the any use of active measures at an unmanned production facility.*

Container ID/ General Condition[14] Aboveground or Buried Tank	Storage Capacity and Type of Oil	Type of Containment/ Drainage Control	Overfill Protection and Testing & Inspections

---

[14] Identify each tank with either an A to indicate aboveground or B for completely buried

ATTACHMENT D: SPCC INSPECTION AND TESTING CHECKLIST
Required Documentation of Tests and Inspections

Records of inspections and tests required by 40 CFR part 112 signed by the appropriate supervisor or inspector must be kept by all facilities with the SPCC Plan for a period of three years. Records of inspections and tests conducted under usual and customary business practices will suffice. Documentation of the following inspections and tests should be kept with the SPCC Plan.

	Inspection or Test	Documentation		Not Applicable
		Present	Not Present	
112.6—Tier I Qualified Facilities				
(a)(3)(iii)	Regular testing of system or documented procedures used instead of liquid level sensing devices specified in §§112.8(c)(8) and 112.12(c)(8) to prevent container overfills	☐	☐	☐
112.7—General SPCC Requirements				
k(2)(i)	Inspection or monitoring of qualified oil-filled operational equipment when the equipment meets the qualification criteria in §112.7(k)(1) and facility owner/operator chooses to implement the alternative requirements in §112.7(k)(2) that include an inspection or monitoring program to detect oil-filled operational equipment failure and discharges	☐	☐	☐
112.8/112.12—Onshore Facilities (excluding oil production facilities) ☐ NA				
(b)(1), (b)(2)	Inspection of storm water released from diked areas into facility drainage directly to a watercourse	☐	☐	☐
(c)(3)	Inspection of rainwater released directly from diked containment areas to a storm drain or open watercourse before release, open and release bypass valve under supervision, and records of drainage events	☐	☐	☐
(c)(4)	Regular leak testing of completely buried metallic storage tanks installed on or after January 10, 1974 and regulated under 40 CFR 112	☐	☐	☐
(c)(6)	Regular integrity testing of aboveground containers and integrity testing after material repairs, including comparison records	☐	☐	☐
(c)(6), (c)(10)	Regular visual inspections of the outsides of aboveground containers, supports and foundations	☐	☐	☐
(c)(6)	Frequent inspections of diked areas for accumulations of oil	☐	☐	☐
(d)(4)	Regular inspections of aboveground valves, piping and appurtenances and assessments of the general condition of flange joints, expansion joints, valve glands and bodies, catch pans, pipeline supports, locking of valves, and metal surfaces	☐	☐	☐
(d)(4)	Integrity and leak testing of buried piping at time of installation, modification, construction, relocation or replacement	☐	☐	☐
112.9—Onshore Oil Production Facilities (excluding drilling and workover facilities) ☐ NA				
(b)(1)	Rainwater released directly from diked containment areas inspected following §§112.8(c)(3)(ii), (iii) and (iv), including records of drainage kept	☐	☐	☐
(b)(2)	Field drainage systems, oil traps, sumps, and skimmers inspected regularly for oil, and accumulations of oil promptly removed	☐	☐	☐
(c)(3)	Containers, foundations and supports inspected visually for deterioration and maintenance needs	☐	☐	☐
(c)(5)(i)	In lieu of having sized secondary containment, flow-through process vessels and associated components visually inspected and/or tested periodically and on a regular schedule for conditions that could result in a discharge as described in §112.1(b)	☐	☐	☐
(d)(1)	All aboveground valves and piping associated with transfers are regularly inspected	☐	☐	☐
(d)(4)(ii)	For flowlines and intra-facility gathering lines without secondary containment, in accordance with §112.7(c), lines are visually inspected and/or tested periodically and on a regular schedule to allow implementing the part 109 contingency plan or the FRP submitted under §112.20	☐	☐	☐

ATTACHMENT E: SPCC CONTINGENCY PLAN REVIEW CHECKLIST ☐
NA

40 CFR Part 109–Criteria for State, Local and Regional Oil Removal Contingency Plans

If SPCC Plan includes an impracticability determination for secondary containment in accordance with §112.7(d), the facility owner/operator is required to provide an oil spill contingency plan following 40 CFR part 109, unless he or she has submitted a FRP under §112.20. An oil spill contingency plan may also be developed, unless the facility owner/operator has submitted a FRP under §112.20 as one of the required alternatives to general secondary containment for qualified oil filled operational equipment in accordance with §112.7(k).

	109.5–Development and implementation criteria for State, local and regional oil removal contingency plans[15]	Yes	No
(a)	Definition of the authorities, responsibilities and duties of all persons, organizations or agencies which are to be involved in planning or directing oil removal operations.	☐	☐
(b)	Establishment of notification procedures for the purpose of early detection and timely notification of an oil discharge including:	☐	☐
(1)	The identification of critical water use areas to facilitate the reporting of and response to oil discharges.	☐	☐
(2)	A current list of names, telephone numbers and addresses of the responsible persons (with alternates) and organizations to be notified when an oil discharge is discovered.	☐	☐
(3)	Provisions for access to a reliable communications system for timely notification of an oil discharge, and the capability of interconnection with the communications systems established under related oil removal contingency plans, particularly State and National plans (e.g., National Contingency Plan (NCP)).	☐	☐
(4)	An established, prearranged procedure for requesting assistance during a major disaster or when the situation exceeds the response capability of the State, local or regional authority.	☐	☐
(c)	Provisions to assure that full resource capability is known and can be committed during an oil discharge situation including:	☐	☐
(1)	The identification and inventory of applicable equipment, materials and supplies which are available locally and regionally.	☐	☐
(2)	An estimate of the equipment, materials and supplies that would be required to remove the maximum oil discharge to be anticipated.	☐	☐
(3)	Development of agreements and arrangements in advance of an oil discharge for the acquisition of equipment, materials and supplies to be used in responding to such a discharge.	☐	☐
(d)	Provisions for well-defined and specific actions to be taken after discovery and notification of an oil discharge including:	☐	☐
(1)	Specification of an oil discharge response operating team consisting of trained, prepared and available operating personnel.	☐	☐
(2)	Pre-designation of a properly qualified oil discharge response coordinator who is charged with the responsibility and delegated commensurate authority for directing and coordinating response operations and who knows how to request assistance from Federal authorities operating under existing national and regional contingency plans.	☐	☐
(3)	A preplanned location for an oil discharge response operations center and a reliable communications system for directing the coordinated overall response operations.	☐	☐
(4)	Provisions for varying degrees of response effort depending on the severity of the oil discharge.	☐	☐
(5)	Specification of the order of priority in which the various water uses are to be protected where more than one water use may be adversely affected as a result of an oil discharge and where response operations may not be adequate to protect all uses.	☐	☐
(e)	Specific and well defined procedures to facilitate recovery of damages and enforcement measures as provided for by State and local statutes and ordinances.	☐	☐

---

[15] The contingency plan should be consistent with all applicable state and local plans, Area Contingency Plans, and the NCP.

TABLE F-1 Tier I Facility Checklist

SUPERVISOR REVIEW/SIGNATURE:	DATE:

SPCC GENERAL APPLICABILITY—40 CFR 112.1

IS THE FACILITY REGULATED UNDER 40 CFR part 112?

The completely buried oil storage capacity is over 42,000 U.S. gallons, **OR** the aggregate aboveground oil storage capacity is over 1,320 U.S. gallons **AND**   ☐ Yes ☐ No

The facility is a non-transportation-related facility engaged in drilling, producing, gathering, storing, processing, refining, transferring, distributing, using, or consuming oil and oil products, which due to its location could reasonably be expected to discharge oil into or upon the navigable waters of the United States   ☐ Yes ☐ No

AFFECTED WATERWAY(S):	DISTANCE:

FLOW PATH TO WATERWAY:

Note: The following storage capacity is not considered in determining applicability of SPCC requirements:

- *Equipment subject to the authority of the U.S. Department of Transportation, U.S. Department of the Interior, or Minerals Management Service, as defined in Memoranda of Understanding dated November 24, 1971, and November 8, 1993; Tank trucks that return to an otherwise regulated facility that contain only residual amounts of oil (EPA Policy letter)*
- *Completely buried tanks subject to all the technical requirements of 40 CFR part 280 or a state program approved under 40 CFR part 281;*
- *Underground oil storage tanks deferred under 40 CFR part 280 that supply emergency diesel generators at a nuclear power generation facility licensed by the Nuclear Regulatory Commission (NRC) and subject to any NRC provision regarding design and quality criteria, including but not limited to CFR part 50;*
- *Any facility or part thereof used exclusively for wastewater treatment (production, recovery or recycling of oil is not considered wastewater treatment); (This does not include other oil containers located at a wastewater treatment facility, such as generator tanks or transformers)*

- *Containers smaller than 55 U.S. gallons;*
- *Permanently closed containers (as defined in §112.2);*
- *Motive power containers(as defined in §112.2);*
- *Hot-mix asphalt or any hot-mix asphalt containers;*
- *Heating oil containers used solely at a single-family residence;*
- *Pesticide application equipment and related mix containers;*
- *Any milk and milk product container and associated piping and appurtenances; and*
- *Intra-facility gathering lines subject to the regulatory requirements of 49 CFR part 192 or 195.*

Does the facility have an SPCC Plan?	☐ Yes ☐ No

FACILITY RESPONSE PLAN (FRP) APPLICABILITY—40 CFR 112.20(f)

A non-transportation related onshore facility is required to prepare and implement an FRP as outlined in 40 CFR 112.20 if:

☐ The facility transfers oil over water to or from vessels and has a total oil storage capacity greater than or equal to 42,000 U.S. gallons, **OR**

☐ The facility has a total oil storage capacity of at least 1 million U.S. gallons, **AND** at least one of the following is true:

    ☐ The facility does not have secondary containment sufficiently large to contain the capacity of the largest aboveground tank plus sufficient freeboard for precipitation.

    ☐ The facility is located at a distance such that a discharge could cause injury to fish and wildlife and sensitive environments.

    ☐ The facility is located such that a discharge would shut down a public drinking water intake.

    ☐ The facility has had a reportable discharge greater than or equal to 10,000 U.S. gallons in the past 5 years.

Facility has FRP: ☐ Yes ☐ No ☐ NA	FRP Number:
Facility has a completed and signed copy of Appendix C, Attachment C-II, "Certification of the Applicability of the Substantial Harm Criteria."	☐ Yes ☐ No

Comments:

SPCC TIER II QUALIFIED FACILITY APPLICABILITY—40 CFR 112.3(g)(2)

The aggregate aboveground oil storage capacity is 10,000 U.S. gallons or less **AND**	☐ Yes ☐ No
In the three years prior to the SPCC Plan self-certification date, or since becoming subject to the rule (if the facility has been in operation for less than three years), the facility has **NOT** had:	
• A single discharge as described in §112.1(b) exceeding 1,000 U.S. gallons, **OR**	☐ Yes ☐ No
• Two discharges as described in §112.1(b) each exceeding 42 U.S. gallons within any twelve-month period[1]	☐ Yes ☐ No

IF *YES* TO ALL OF THE ABOVE, THEN THE FACILITY IS A TIER II QUALIFIED FACILITY[2]
SEE ATTACHMENT D FOR TIER II QUALIFIED FACILITY CHECKLIST

REQUIREMENTS FOR PREPARATION AND IMPLEMENTATION OF A SPCC PLAN—40 CFR 112.3

Date facility began operations:

Date of initial SPCC Plan preparation:	Current Plan version (date/number):

112.3(a)	**For facilities (except farms), including mobile or portable facilities:**	
	• In operation on or prior to November 10, 2011: Plan prepared and/or amended and fully implemented by **November 10, 2011**	☐ Yes ☐ No ☐ NA
	• Beginning operations after November 10, 2011, Plan prepared and fully implemented before beginning operations	☐ Yes ☐ No ☐ NA
	For farms (as defined in §112.2):	
	• In operation on or prior to August 16, 2002: Plan maintained, amended and implemented by **May 10, 2013**	☐ Yes ☐ No ☐ NA
	• Beginning operations after August 16, 2002 through May 10, 2013: Plan prepared and fully implemented by **May 10, 2013**	☐ Yes ☐ No ☐ NA
	• Beginning operations after May 10, 2013: Plan prepared and fully implemented before beginning operations	☐ Yes ☐ No ☐ NA
112.3(d)	Plan is certified by a registered Professional Engineer (PE) and includes statements that the PE attests:	☐ Yes ☐ No ☐ NA
	• PE is familiar with the requirements of 40 CFR part 112	☐ Yes ☐ No ☐ NA
	• PE or agent has visited and examined the facility	☐ Yes ☐ No ☐ NA
	• Plan is prepared in accordance with good engineering practice including consideration of applicable industry standards and the requirements of 40 CFR part 112	☐ Yes ☐ No ☐ NA
	• Procedures for required inspections and testing have been established	☐ Yes ☐ No ☐ NA
	• Plan is adequate for the facility	☐ Yes ☐ No ☐ NA
PE Name:	License No.:	State:	Date of certification:
112.3(e)(1)	Plan is available onsite if attended at least 4 hours per day. If facility is unattended, Plan is available at the nearest field office. *(Please note nearest field office contact information in comments section below.)*	☐ Yes ☐ No ☐ NA

---

[1] Oil discharges that result from natural disasters, acts of war, or terrorism are not included in this determination. The gallon amount(s) specified (either 1,000 or 42) refers to the amount of oil that actually reaches navigable waters or adjoining shorelines not the total amount of oil spilled. The entire volume of the discharge is oil for this determination.
[2] An owner/operator who self-certifies a Tier II SPCC Plan may include environmentally equivalent alternatives and/or secondary containment impracticability determinations when reviewed and certified by a PE.

Comments:	

AMENDMENT OF SPCC PLAN BY REGIONAL ADMINISTRATOR (RA)—40 CFR 112.4

112.4(a),(c)	Has the facility discharged more than 1,000 U.S. gallons of oil in a single reportable discharge or more than 42 U.S. gallons in each of two reportable discharges in any 12-month period?[3]	☐ Yes ☐ No
If **YES**	• Was information submitted to the RA as required in §112.4(a)?[4]	☐ Yes ☐ No ☐ NA
	• Was information submitted to the appropriate agency or agencies in charge of oil pollution control activities in the State in which the facility is located§112.4(c)	☐ Yes ☐ No ☐ NA
	• Date(s) and volume(s) of reportable discharges(s) under this section: _____	
	• Were the discharges reported to the NRC[5]?	☐ Yes ☐ No
112.4(d),(e)	Have changes required by the RA been implemented in the Plan and/or facility?	☐ Yes ☐ No ☐ NA

Comments:	

AMENDMENT OF SPCC PLAN BY THE OWNER OR OPERATOR—40 CFR 112.5

112.5(a)	Has there been a change at the facility that materially affects the potential for a discharge described in §112.1(b)?	☐ Yes ☐ No
If **YES**	• Was the Plan amended within six months of the change?	☐ Yes ☐ No
	• Were amendments implemented within six months of any Plan amendment?	☐ Yes ☐ No
112.5(b)	Review and evaluation of the Plan completed at least once every 5 years?	☐ Yes ☐ No ☐ NA
	Following Plan review, was Plan amended within six months to include more effective prevention and control technology that has been field-proven to significantly reduce the likelihood of a discharge described in §112.1(b)?	☐ Yes ☐ No ☐ NA
	Amendments implemented within six months of any Plan amendment?	☐ Yes ☐ No ☐ NA
	Five year Plan review and evaluation documented?	☐ Yes ☐ No ☐ NA
112.5(c)	Professional Engineer certification of any technical Plan amendments in accordance with all applicable requirements of §112.3(d) *[Except for self-certified Plans]*	☐ Yes ☐ No ☐ NA

Name:	License No.:	State:	Date of certification:

Reason for amendment:	

---

[3] A reportable discharge is a discharge as described in §112.1(b)(see 40 CFR part 110). The gallon amount(s) specified (either 1,000 or 42) refers to the amount of oil that actually reaches navigable waters or adjoining shorelines not the total amount of oil spilled. The entire volume of the discharge is oil for this determination.

[4] Triggering this threshold may disqualify the facility from meeting the Qualified Facility criteria if it occurred in the three years prior to self certification

[5] Inspector Note-Confirm any spills identified above were reported to NRC

Comments:

GENERAL SPCC REQUIREMENTS—40 CFR 112.7		PLAN	FIELD
Management approval at a level of authority to commit the necessary resources to fully implement the Plan[6]		☐ Yes ☐ No	
Plan follows sequence of the rule or is an equivalent Plan meeting all applicable rule requirements and includes a cross-reference of provisions		☐ Yes ☐ No ☐ NA	
If Plan calls for facilities, procedures, methods, or equipment not yet fully operational, details of their installation and start-up are discussed *(Note: Relevant for inspection evaluation and testing baselines.)*		☐ Yes ☐ No ☐ NA	
112.7(a)(2)	The Plan includes deviations from the requirements of §§112.7(g), (h)(2) and (3), and (i) and applicable subparts B and C of the rule, except the secondary containment requirements in §§112.7(c) and (h)(1), 112.8(c)(2),112.8(c)(11), 112.12(c)(2), and 112.12(c)(11)	☐ Yes ☐ No ☐ NA	
If **YES**	• The Plan states reasons for nonconformance	☐ Yes ☐ No ☐ NA	
	• Alternative measures described in detail and provide equivalent environmental protection *(Note: Inspector should document if the environmental equivalence is implemented in the field, in accordance with the Plan's description)*	☐ Yes ☐ No ☐ NA	☐ Yes ☐ No ☐ NA

Describe each deviation and reasons for nonconformance:

---

[6] May be part of the Plan or demonstrated elsewhere.

		PLAN	FIELD
112.7(a)(3)	Plan describes physical layout of facility and includes a diagram[7] that identifies: • Location and contents of all regulated fixed oil storage containers • Storage areas where mobile or portable containers are located • Completely buried tanks otherwise exempt from the SPCC requirements (marked as "exempt") • Transfer stations • Connecting pipes, including intra-facility gathering lines that are otherwise exempt from the requirements of this part under §112.1(d)(11)	☐ Yes ☐ No	☐ Yes ☐ No
	Plan addresses each of the following:		
(i)	For each fixed container, type of oil and storage capacity (see Attachment A of this checklist). For mobile or portable containers, type of oil and storage capacity for each container or an estimate of the potential number of mobile or portable containers, the types of oil, and anticipated storage capacities	☐ Yes ☐ No	☐ Yes ☐ No
(ii)	Discharge prevention measures, including procedures for routine handling of products (loading, unloading, and facility transfers, etc.)	☐ Yes ☐ No	☐ Yes ☐ No
(iii)	Discharge or drainage controls, such as secondary containment around containers, and other structures, equipment, and procedures for the control of a discharge	☐ Yes ☐ No	☐ Yes ☐ No
(iv)	Countermeasures for discharge discovery, response, and cleanup (both facility's and contractor's resources)	☐ Yes ☐ No	☐ Yes ☐ No
(v)	Methods of disposal of recovered materials in accordance with applicable legal requirements	☐ Yes ☐ No	
(vi)	Contact list and phone numbers for the facility response coordinator, National Response Center, cleanup contractors with an agreement for response, and all Federal, State, and local agencies who must be contacted in the case of a discharge as described in §112.1(b)	☐ Yes ☐ No	
112.7(a)(4)	**Does not apply if the facility has submitted an FRP under §112.20:** Plan includes information and procedures that enable a person reporting an oil discharge as described in §112.1(b) to relate information on the: • Exact address or location and phone number of the facility; • Date and time of the discharge; • Type of material discharged; • Estimates of the total quantity discharged; • Estimates of the quantity discharged as described in §112.1(b); • Source of the discharge; • Description of all affected media; • Cause of the discharge; • Damages or injuries caused by the discharge; • Actions being used to stop, remove, and mitigate the effects of the discharge; • Whether an evacuation may be needed; and • Names of individuals and/or organizations who have also been contacted.	☐ Yes ☐ No ☐ NA	
112.7(a)(5)	**Does not apply if the facility has submitted a FRP under §112.20:** Plan organized so that portions describing procedures to be used when a discharge occurs will be readily usable in an emergency	☐ Yes ☐ No ☐ NA	
112.7(b)	Plan includes a prediction of the direction, rate of flow, and total quantity of oil that could be discharged for each type of major equipment failure where experience indicates a reasonable potential for equipment failure	☐ Yes ☐ No ☐ NA	
Comments:			

---

[7] Note in comments any discrepancies between the facility diagram, the description of the physical layout of facility, and what is observed in the field

		PLAN	FIELD
112.7(c)	Appropriate containment and/or diversionary structures or equipment are provided to prevent a discharge as described in §112.1(b), **except as provided in §112.7(k) of this section for certain qualified operational equipment**. The entire containment system, including walls and floors, are capable of containing oil and are constructed to prevent escape of a discharge from the containment system before cleanup occurs. The method, design, and capacity for secondary containment address the typical failure mode and the most likely quantity of oil that would be discharged. See Attachment A of this checklist.		

For onshore facilities, one of the following or its equivalent:

- Dikes, berms, or retaining walls sufficiently impervious to contain oil;
- Curbing or drip pans;
- Sumps and collection systems;
- Culverting, gutters or other drainage systems;
- Weirs, booms or other barriers;
- Spill diversion pond;
- Retention ponds; or
- Sorbent materials.

Identify which of the following are present at the facility and if appropriate containment and/or diversionary structures or equipment are provided as described above:

	PLAN	FIELD
☐ Bulk storage containers	☐ Yes ☐ No ☐ NA	☐ Yes ☐ No ☐ NA
☐ Mobile/portable containers	☐ Yes ☐ No ☐ NA	☐ Yes ☐ No ☐ NA
☐ Oil-filled operational equipment (as defined in 112.2)	☐ Yes ☐ No ☐ NA	☐ Yes ☐ No ☐ NA
☐ Other oil-filled equipment (i.e., manufacturing equipment)	☐ Yes ☐ No ☐ NA	☐ Yes ☐ No ☐ NA
☐ Piping and related appurtenances	☐ Yes ☐ No ☐ NA	☐ Yes ☐ No ☐ NA
☐ Mobile refuelers or non-transportation-related tank cars	☐ Yes ☐ No ☐ NA	☐ Yes ☐ No ☐ NA
☐ Transfer areas, equipment and activities	☐ Yes ☐ No ☐ NA	☐ Yes ☐ No ☐ NA
☐ Identify any other equipment or activities that are not listed above: _____	☐ Yes ☐ No ☐ NA	☐ Yes ☐ No ☐ NA

		PLAN	FIELD
112.7(d)	Secondary containment for one (or more) of the following provisions is determined to be impracticable: ☐ General secondary containment §112.7(c) ☐ Loading/unloading rack §112.7(h)(1) ☐ Bulk storage containers §§112.8(c)(2)/112.12(c)(2) ☐ Mobile/portable containers §§112.8(c)(11)/112.12(c)(11)	☐ Yes ☐ No	
If YES	• The impracticability of secondary containment is clearly demonstrated and described in the Plan	☐ Yes ☐ No ☐ NA	☐ Yes ☐ No ☐ NA
	• For bulk storage containers,[8] periodic integrity testing of containers and integrity and leak testing of the associated valves and piping is conducted	☐ Yes ☐ No ☐ NA	☐ Yes ☐ No ☐ NA
	(Does not apply if the facility has submitted a FRP under §112.20): • Contingency Plan following the provisions of 40 CFR part 109 is provided (see Attachment C of this checklist) **AND**	☐ Yes ☐ No ☐ NA	
	• Written commitment of manpower, equipment, and materials required to expeditiously control and remove any quantity of oil discharged that may be harmful	☐ Yes ☐ No ☐ NA	☐ Yes ☐ No ☐ NA

Comments:

---

[8] These additional requirements apply only to bulk storage containers, when an impracticability determination has been made by the PE

		PLAN	FIELD
112.7(e)	Inspections and tests conducted in accordance with written procedures	☐ Yes ☐ No	☐ Yes ☐ No
	Record of inspections or tests signed by supervisor or inspector	☐ Yes ☐ No	☐ Yes ☐ No
	Kept with Plan for at least 3 years (see Attachment B of this checklist)[9]	☐ Yes ☐ No	☐ Yes ☐ No
112.7(f)	Personnel, training, and oil discharge prevention procedures		
(1)	Training of oil-handling personnel in operation and maintenance of equipment to prevent discharges; discharge procedure protocols; applicable pollution control laws, rules, and regulations; general facility operations; and contents of SPCC Plan	☐ Yes ☐ No ☐ NA	☐ Yes ☐ No ☐ NA
(2)	Person designated as accountable for discharge prevention at the facility and reports to facility management	☐ Yes ☐ No ☐ NA	☐ Yes ☐ No ☐ NA
(3)	Discharge prevention briefings conducted at least once a year for oil handling personnel to assure adequate understanding of the Plan. Briefings highlight and describe known discharges as described in §112.1(b) or failures, malfunctioning components, and any recently developed precautionary measures	☐ Yes ☐ No ☐ NA	☐ Yes ☐ No ☐ NA
112.7(g)	Plan describes how to: • Secure and control access to the oil handling, processing and storage areas; • Secure master flow and drain valves; • Prevent unauthorized access to starter controls on oil pumps; • Secure out-of-service and loading/unloading connections of oil pipelines; and • Address the appropriateness of security lighting to both prevent acts of vandalism and assist in the discovery of oil discharges.	☐ Yes ☐ No ☐ NA	☐ Yes ☐ No ☐ NA
112.7(h)	Tank car and tank truck loading/unloading rack[10] is present at the facility		☐ Yes ☐ No
	Loading/unloading rack means a fixed structure (such as a platform, gangway) necessary for loading or unloading a tank truck or tank car, which is located at a facility subject to the requirements of this part. A loading/unloading rack includes a loading or unloading arm, and may include any combination of the following: piping assemblages, valves, pumps, shut-off devices, overfill sensors, or personnel safety devices.		
If **YES** (1)	Does loading/unloading rack drainage flow to catchment basin or treatment facility designed to handle discharges or use a quick drainage system?	☐ Yes ☐ No ☐ NA	☐ Yes ☐ No ☐ NA
	Containment system holds at least the maximum capacity of the largest single compartment of a tank car/truck loaded/unloaded at the facility	☐ Yes ☐ No ☐ NA	☐ Yes ☐ No ☐ NA
(2)	An interlocked warning light or physical barriers, warning signs, wheel chocks, or vehicle brake interlock system in the area adjacent to the **loading or unloading rack** to prevent vehicles from departing before complete disconnection of flexible or fixed oil transfer lines	☐ Yes ☐ No ☐ NA	☐ Yes ☐ No ☐ NA
(3)	Lower-most drains and all outlets on tank cars/trucks inspected prior to filling/departure, and, if necessary ensure that they are tightened, adjusted, or replaced to prevent liquid discharge while in transit	☐ Yes ☐ No ☐ NA	☐ Yes ☐ No ☐ NA

Comments:

---

[9] Records of inspections and tests kept under usual and customary business practices will suffice
[10] Note that a tank car/truck loading/unloading rack must be present for §112.7(h) to apply

		PLAN	FIELD
112.7(i)	Brittle fracture evaluation of field-constructed aboveground containers is conducted after tank repair, alteration, reconstruction, or change in service that might affect the risk of a discharge or after a discharge/failure due to brittle fracture or other catastrophe, and appropriate action taken as necessary (applies to only field-constructed aboveground containers)	☐ Yes ☐ No ☐ NA	☐ Yes ☐ No ☐ NA
112.7(j)	Discussion of conformance with applicable more stringent State rules, regulations, and guidelines and other effective discharge prevention and containment procedures listed in 40 CFR part 112	☐ Yes ☐ No ☐ NA	

112.7(k)
 	
If YES
	Qualified oil-filled operational equipment is present at the facility[11]
	
	Oil-filled operational equipment means equipment that includes an oil storage container (or multiple containers) in which the oil is present solely to support the function of the apparatus or the device. Oil-filled operational equipment is not considered a bulk storage container, and does not include oil-filled manufacturing equipment (flow-through process). Examples of oil-filled operational equipment include, but are not limited to, hydraulic systems, lubricating systems (e.g. , those for pumps, compressors and other rotating equipment, including pumpjack lubrication systems), gear boxes, machining coolant systems, heat transfer systems, transformers, circuit breakers, electrical switches, and other systems containing oil solely to enable the operation of the device.
	Check which apply:
	
	  Secondary Containment provided in accordance with 112.7(c) ☐
	  Alternative measure described below (confirm eligibility) ☐
		☐ Yes ☐ No	

112.7(k)	Qualified Oil-Filled Operational Equipment • Has a single reportable discharge as described in §112.1(b) from any oil-filled operational equipment exceeding 1,000 U.S. gallons occurred within the three years prior to Plan certification date?	☐ Yes ☐ No ☐ NA	
	• Have two reportable discharges as described in §112.1(b) from any oil-filled operational equipment each exceeding 42 U.S. gallons occurred within any 12-month period within the three years prior to Plan certification date?[12]	☐ Yes ☐ No ☐ NA	
	If YES for either, secondary containment in accordance with §112.7(c) is required		
	• Facility procedure for inspections or monitoring program to detect equipment failure and/or a discharge is established and documented ***Does not apply if the facility has submitted a FRP under §112.20:*** • Contingency plan following 40 CFR part 109 (see Attachment C of this checklist) is provided in Plan **AND** • Written commitment of manpower, equipment, and materials required to expeditiously control and remove any quantity of oil discharged that may be harmful is provided in Plan	☐ Yes ☐ No ☐ NA ☐ Yes ☐ No ☐ NA ☐ Yes ☐ No ☐ NA	☐ Yes ☐ No ☐ NA

Comments:

---

[11] This provision does not apply to oil-filled manufacturing equipment (flow-through process)

[12] Oil discharges that result from natural disasters, acts of war, or terrorism are not included in this determination. The gallon amount(s) specified (either 1,000 or 42) refers to the amount of oil that actually reaches navigable waters or adjoining shorelines not the total amount of oil spilled. The entire volume of the discharge is oil for this determination.

ONSHORE FACILITIES (EXCLUDING PRODUCTION) 40 CFR 112.8/112.12	PLAN	FIELD
112.8(b)/ 112.12(b) Facility Drainage		
Diked Areas (1) Drainage from diked storage areas is: • Restrained by valves, except where facility systems are designed to control such discharge, **OR** • Manually activated pumps or ejectors are used and the condition of the accumulation is inspected prior to draining dike to ensure no oil will be discharged	☐ Yes ☐ No ☐ NA	☐ Yes ☐ No ☐ NA
(2) Diked storage area drain valves are manual, open-and-closed design (not flapper-type drain valves)	☐ Yes ☐ No ☐ NA	☐ Yes ☐ No ☐ NA
If drainage is released directly to a watercourse and not into an onsite wastewater treatment plant, retained storm water is inspected and discharged per §§112.8(c)(3)(ii), (iii), and (iv) or §§112.12(c)(3)(ii), (iii), and (iv).	☐ Yes ☐ No ☐ NA	☐ Yes ☐ No ☐ NA
Undiked Areas (3) Drainage from undiked areas with a potential for discharge designed to flow into ponds, lagoons, or catchment basins to retain oil or return it to facility. Catchment basin located away from flood areas.[13]	☐ Yes ☐ No ☐ NA	☐ Yes ☐ No ☐ NA
(4) If facility drainage not engineered as in (b)(3) (i.e., drainage flows into ponds, lagoons, or catchment basins) then the facility is equipped with a diversion system to retain oil in the facility in the event of an uncontrolled discharge.[14]	☐ Yes ☐ No ☐ NA	☐ Yes ☐ No ☐ NA
(5) Are facility drainage waters continuously treated in more than one treatment unit and pump transfer is needed?	☐ Yes ☐ No ☐ NA	☐ Yes ☐ No ☐ NA
If YES • Two "lift" pumps available and at least one permanently installed	☐ Yes ☐ No ☐ NA	☐ Yes ☐ No ☐ NA
• Facility drainage systems engineered to prevent a discharge as described in §112.1(b) in the case of equipment failure or human error	☐ Yes ☐ No ☐ NA	☐ Yes ☐ No ☐ NA

Comments:

112.8(c)/112.12(c) Bulk Storage Containers ☐ NA

Bulk storage container means any container used to store oil. These containers are used for purposes including, but not limited to, the storage of oil prior to use, while being used, or prior to further distribution in commerce. Oil-filled electrical, operating, or manufacturing equipment is not a bulk storage container.

If bulk storage containers are not present, mark this section Not Applicable (NA). If present, complete this section and Attachment A of this checklist.

	PLAN	FIELD
(1) Containers materials and construction are compatible with material stored and conditions of storage such as pressure and temperature	☐ Yes ☐ No ☐ NA	☐ Yes ☐ No ☐ NA
(2) Except for mobile refuelers and other non-transportation-related tank trucks, construct all bulk storage tank installations with secondary containment to hold capacity of largest container and sufficient freeboard for precipitation	☐ Yes ☐ No ☐ NA	☐ Yes ☐ No ☐ NA
Diked areas sufficiently impervious to contain discharged oil **OR**	☐ Yes ☐ No ☐ NA	☐ Yes ☐ No ☐ NA
Alternatively, any discharge to a drainage trench system will be safely confined in a facility catchment basin or holding pond	☐ Yes ☐ No ☐ NA	☐ Yes ☐ No ☐ NA

---

[13] Oil discharges that result from natural disasters, acts of war, or terrorism are not included in this determination. The gallon amount(s) specified (either 1,000 or 42) refers to the amount of oil that actually reaches navigable waters or adjoining shorelines not the total amount of oil spilled. The entire volume of the discharge is oil for this determination.
[14] These provisions apply only when a facility drainage system is used for containment; otherwise mark NA

		PLAN	FIELD
(3)	Is there drainage of uncontaminated rainwater from diked areas into a storm drain or open watercourse?	☐ Yes ☐ No ☐ NA	☐ Yes ☐ No ☐ NA
If **YES**	• Bypass valve normally sealed closed	☐ Yes ☐ No ☐ NA	☐ Yes ☐ No ☐ NA
	• Retained rainwater is inspected to ensure that its presence will not cause a discharge as described in §112.1(b)	☐ Yes ☐ No ☐ NA	☐ Yes ☐ No ☐ NA
	• Bypass valve opened and resealed under responsible supervision	☐ Yes ☐ No ☐ NA	☐ Yes ☐ No ☐ NA
	• Adequate records of drainage are kept; for example, records required under permits issued in accordance with 40 CFR §§122.41(j)(2) and (m)(3)	☐ Yes ☐ No ☐ NA	☐ Yes ☐ No ☐ NA
(4)	For completely buried metallic tanks installed on or after January 10, 1974 (if not exempt from SPCC regulation because subject to all of the technical requirements of 40 CFR part 280 or 281):		
	• Provide corrosion protection with coatings or cathodic protection compatible with local soil conditions	☐ Yes ☐ No ☐ NA	☐ Yes ☐ No ☐ NA
	• Regular leak testing conducted	☐ Yes ☐ No ☐ NA	☐ Yes ☐ No ☐ NA
(5)	The buried section of partially buried or bunkered metallic tanks protected from corrosion with coatings or cathodic protection compatible with local soil conditions	☐ Yes ☐ No ☐ NA	☐ Yes ☐ No ☐ NA
(6)	• Test or inspect each aboveground container for integrity on a regular schedule and whenever you make material repairs. Techniques include, but are not limited to: visual inspection, hydrostatic testing, radiographic testing, ultrasonic testing, acoustic emissions testing, or other system of non-destructive testing	☐ Yes ☐ No ☐ NA	☐ Yes ☐ No ☐ NA
	• Appropriate qualifications for personnel performing tests and inspections are identified in the Plan and have been assessed in accordance with industry standards	☐ Yes ☐ No ☐ NA	☐ Yes ☐ No ☐ NA
	• The frequency and type of testing and inspections are documented, are in accordance with industry standards and take into account the container size, configuration and design	☐ Yes ☐ No ☐ NA	☐ Yes ☐ No ☐ NA
	• Comparison records of aboveground container integrity testing are maintained	☐ Yes ☐ No ☐ NA	☐ Yes ☐ No ☐ NA
	• Container supports and foundations regularly inspected	☐ Yes ☐ No ☐ NA	☐ Yes ☐ No ☐ NA
	• Outside of containers frequently inspected for signs of deterioration, discharges, or accumulation of oil inside diked areas	☐ Yes ☐ No ☐ NA	☐ Yes ☐ No ☐ NA
	• Records of all inspections and tests maintained[15]	☐ Yes ☐ No ☐ NA	☐ Yes ☐ No ☐ NA
Integrity Testing Standard identified in the Plan:			
112.12 (c)(6)(ii) *(Applies to AFVO Facilities only)*	Conduct formal visual inspection on a regular schedule for bulk storage containers that meet all of the following conditions: • Subject to 21 CFR part 110; • Have no external insulation; and • Elevated; • Shop-fabricated. • Constructed of austenitic stainless steel;	☐ Yes ☐ No ☐ NA	☐ Yes ☐ No ☐ NA
	In addition, you must frequently inspect the outside of the container for signs of deterioration, discharges, or accumulation of oil inside diked areas.	☐ Yes ☐ No ☐ NA	☐ Yes ☐ No ☐ NA
	You must determine and document in the Plan the appropriate qualifications for personnel performing tests and inspections.[16]	☐ Yes ☐ No ☐ NA	☐ Yes ☐ No ☐ NA

---

[15] Records of inspections and tests kept under usual and customary business practices will suffice

		PLAN	FIELD
(7)	Leakage through defective internal heating coils controlled:		
	• Steam returns and exhaust lines from internal heating coils that discharge into an open watercourse are monitored for contamination, **OR**	☐ Yes ☐ No ☐ NA	☐ Yes ☐ No ☐ NA
	• Steam returns and exhaust lines pass through a settling tank, skimmer, or other separation or retention system	☐ Yes ☐ No ☐ NA	☐ Yes ☐ No ☐ NA
(8)	Each container is equipped with at least one of the following for liquid level sensing:	☐ Yes ☐ No ☐ NA	☐ Yes ☐ No ☐ NA
	• High liquid level alarms with an audible or visual signal at a constantly attended operation or surveillance station, or audible air vent in smaller facilities; • High liquid level pump cutoff devices set to stop flow at a predetermined container content level; • Direct audible or code signal communication between container gauger and pumping station; • Fast response system for determining liquid level (such as digital computers, telepulse, or direct vision gauges) and a person present to monitor gauges and overall filling of bulk containers; or • Regularly test liquid level sensing devices to ensure proper operation.		
(9)	Effluent treatment facilities observed frequently enough to detect possible system upsets that could cause a discharge as described in §112.1(b)	☐ Yes ☐ No ☐ NA	☐ Yes ☐ No ☐ NA
(10)	Visible discharges which result in a loss of oil from the container, including but not limited to seams, gaskets, piping, pumps, valves, rivets, and bolts are promptly corrected and oil in diked areas is promptly removed	☐ Yes ☐ No ☐ NA	☐ Yes ☐ No ☐ NA
(11)	Mobile or portable containers positioned to prevent a discharge as described in §112.1(b).	☐ Yes ☐ No ☐ NA	☐ Yes ☐ No ☐ NA
	Mobile or portable containers (excluding mobile refuelers and other non-transportation-related tank trucks) have secondary containment with sufficient capacity to contain the largest single compartment or container and sufficient freeboard to contain precipitation	☐ Yes ☐ No ☐ NA	☐ Yes ☐ No ☐ NA
112.8(d)/112.12(d)Facility transfer operations, pumping, and facility process			
(1)	Buried piping installed or replaced on or after August 16, 2002 has protective wrapping or coating	☐ Yes ☐ No ☐ NA	☐ Yes ☐ No ☐ NA
	Buried piping installed or replaced on or after August 16, 2002 is also cathodically protected or otherwise satisfies corrosion protection standards for piping in 40 CFR part 280 or 281	☐ Yes ☐ No ☐ NA	☐ Yes ☐ No ☐ NA
	Buried piping exposed for any reason is inspected for deterioration; corrosion damage is examined; and corrective action is taken	☐ Yes ☐ No ☐ NA	☐ Yes ☐ No ☐ NA
(2)	Piping terminal connection at the transfer point is marked as to origin and capped or blank-flanged when not in service or in standby service for an extended time	☐ Yes ☐ No ☐ NA	☐ Yes ☐ No ☐ NA
(3)	Pipe supports are properly designed to minimize abrasion and corrosion and allow for expansion and contraction	☐ Yes ☐ No ☐ NA	☐ Yes ☐ No ☐ NA
(4)	Aboveground valves, piping, and appurtenances such as flange joints, expansion joints, valve glands and bodies, catch pans, pipeline supports, locking of valves, and metal surfaces are inspected regularly to assess their general condition	☐ Yes ☐ No ☐ NA	☐ Yes ☐ No ☐ NA
	Integrity and leak testing conducted on buried piping at time of installation, modification, construction, relocation, or replacement	☐ Yes ☐ No ☐ NA	☐ Yes ☐ No ☐ NA
(5)	Vehicles warned so that no vehicle endangers aboveground piping and other oil transfer operations	☐ Yes ☐ No ☐ NA	☐ Yes ☐ No ☐ NA

Comments:

ATTACHMENT A: SPCC FIELD INSPECTION AND PLAN REVIEW TABLE
Documentation of Field Observations for Containers and Associated Requirements

Inspectors should use this table to document observations of containers as needed.

<u>Containers and Piping</u>

Check containers for leaks, specifically looking for: drip marks, discoloration of tanks, puddles containing spilled or leaked material, corrosion, cracks, and localized dead vegetation, and standards/specifications of construction.

Check aboveground container foundation for: cracks, discoloration, and puddles containing spilled or leaked material, settling, gaps between container and foundation, and damage caused by vegetation roots.

Check all piping for: droplets of stored material, discoloration, corrosion, bowing of pipe between supports, evidence of stored material seepage from valves or seals, evidence of leaks, and localized dead vegetation. For all aboveground piping, include the general condition of flange joints, valve glands and bodies, drip pans, pipe supports, bleeder and gauge valves, and other such items (Document in comments section of §112.8(d) or 112.12(d).)

<u>Secondary Containment</u> **(Active and Passive)**

Check secondary containment for: containment system (including walls and floor) ability to contain oil such that oil will not escape the containment system before cleanup occurs, proper sizing, cracks, discoloration, presence of spilled or leaked material (standing liquid), erosion, corrosion, penetrations in the containment system, and valve conditions.

Check dike or berm systems for: level of precipitation in dike/available capacity, operational status of drainage valves (closed), dike or berm impermeability, debris, erosion, impermeability of the earthen floor/walls of diked area, and location/status of pipes, inlets, drainage around and beneath containers, presence of oil discharges within diked areas.

Check drainage systems for: an accumulation of oil that may have resulted from any small discharge, including field drainage systems (such as drainage ditches or road ditches), and oil traps, sumps, or skimmers. Ensure any accumulations of oil have been promptly removed.

Check retention and drainage ponds for: erosion, available capacity, presence of spilled or leaked material, debris, and stressed vegetation.

Check active measures (countermeasures) for: amount indicated in plan is available and appropriate; deployment procedures are realistic; material is located so that they are readily available; efficacy of discharge detection; availability of personnel and training, appropriateness of measures to prevent a discharge as described in §112.1(b).

Container ID/ General Condition[16] Aboveground or Buried Tank	Storage Capacity and Type of Oil	Type of Containment/ Drainage Control	Overfill Protection and Testing & Inspections

---

[16] Identify each tank with either an A to indicate aboveground or B for completely buried

ATTACHMENT B: SPCC INSPECTION AND TESTING CHECKLIST
Required Documentation of Tests and Inspections

Records of inspections and tests required by 40 CFR part 112 signed by the appropriate supervisor or inspector must be kept by all facilities with the SPCC Plan for a period of three years. Records of inspections and tests conducted under usual and customary business practices will suffice. Documentation of the following inspections and tests should be kept with the SPCC Plan.

Inspection or Test	Documentation		Not Applicable
	Present	Not Present	
112.7–General SPCC Requirements			
(d) Integrity testing for bulk storage containers with no secondary containment system and for which an impracticability determination has been made	☐	☐	☐
(d) Integrity and leak testing of valves and piping associated with bulk storage containers with no secondary containment system and for which an impracticability determination has been made	☐	☐	☐
(h)(3) Inspection of lowermost drain and all outlets of tank car or tank truck prior to filling and departure from loading/unloading rack	☐	☐	☐
(i) Evaluation of field-constructed aboveground containers for potential for brittle fracture or other catastrophic failure when the container undergoes a repair, alteration, reconstruction or change in service or has discharged oil or failed due to brittle fracture failure or other catastrophe	☐	☐	☐
k(2)(i) Inspection or monitoring of qualified oil-filled operational equipment when the equipment meets the qualification criteria in §112.7(k)(1) and facility owner/operator chooses to implement the alternative requirements in §112.7(k)(2) that include an inspection or monitoring program to detect oil-filled operational equipment failure and discharges	☐	☐	☐
112.8/112.12–Onshore Facilities (excluding oil production facilities)			
(b)(1), (b)(2) Inspection of storm water released from diked areas into facility drainage directly to a watercourse	☐	☐	☐
(c)(3) Inspection of rainwater released directly from diked containment areas to a storm drain or open watercourse before release, open and release bypass valve under supervision, and records of drainage events	☐	☐	☐
(c)(4) Regular leak testing of completely buried metallic storage tanks installed on or after January 10, 1974 and regulated under 40 CFR 112	☐	☐	☐
(c)(6) Regular integrity testing of aboveground containers and integrity testing after material repairs, including comparison records	☐	☐	☐
(c)(6), (c)(10) Regular visual inspections of the outsides of aboveground containers, supports and foundations	☐	☐	☐
(c)(6) Frequent inspections of diked areas for accumulations of oil	☐	☐	☐
(c)(8)(v) Regular testing of liquid level sensing devices to ensure proper operation	☐	☐	☐
(c)(9) Frequent observations of effluent treatment facilities to detect possible system upsets that could cause a discharge as described in §112.1(b)	☐	☐	☐
(d)(1) Inspection of buried piping for damage when piping is exposed and additional examination of corrosion damage and corrective action, if present	☐	☐	☐
(d)(4) Regular inspections of aboveground valves, piping and appurtenances and assessments of the general condition of flange joints, expansion joints, valve glands and bodies, catch pans, pipeline supports, locking of valves, and metal surfaces	☐	☐	☐
(d)(4) Integrity and leak testing of buried piping at time of installation, modification, construction, relocation or replacement	☐	☐	☐

ATTACHMENT C: SPCC CONTINGENCY PLAN REVIEW CHECKLIST ☐ NA
40 CFR Part 109–Criteria for State, Local and Regional Oil Removal Contingency Plans

If SPCC Plan includes an impracticability determination for secondary containment in accordance with §112.7(d), the facility owner/operator is required to provide an oil spill contingency plan following 40 CFR part 109, unless he or she has submitted a FRP under §112.20. An oil spill contingency plan may also be developed, unless the facility owner/operator has submitted a FRP under §112.20 as one of the required alternatives to general secondary containment for qualified oil filled operational equipment in accordance with §112.7(k).

109.5–Development and implementation criteria for State, local and regional oil removal contingency plans[17]	Yes	No
(a) Definition of the authorities, responsibilities and duties of all persons, organizations or agencies which are to be involved in planning or directing oil removal operations.	☐	☐
(b) Establishment of notification procedures for the purpose of early detection and timely notification of an oil discharge including:	☐	☐
(1) The identification of critical water use areas to facilitate the reporting of and response to oil discharges.	☐	☐
(2) A current list of names, telephone numbers and addresses of the responsible persons (with alternates) and organizations to be notified when an oil discharge is discovered.	☐	☐
(3) Provisions for access to a reliable communications system for timely notification of an oil discharge, and the capability of interconnection with the communications systems established under related oil removal contingency plans, particularly State and National plans (e.g., National Contingency Plan (NCP)).	☐	☐
(4) An established, prearranged procedure for requesting assistance during a major disaster or when the situation exceeds the response capability of the State, local or regional authority.	☐	☐
(c) Provisions to assure that full resource capability is known and can be committed during an oil discharge situation including:	☐	☐
(1) The identification and inventory of applicable equipment, materials and supplies which are available locally and regionally.	☐	☐
(2) An estimate of the equipment, materials and supplies that would be required to remove the maximum oil discharge to be anticipated.	☐	☐
(3) Development of agreements and arrangements in advance of an oil discharge for the acquisition of equipment, materials and supplies to be used in responding to such a discharge.	☐	☐
(d) Provisions for well-defined and specific actions to be taken after discovery and notification of an oil discharge including:	☐	☐
(1) Specification of an oil discharge response operating team consisting of trained, prepared and available operating personnel.	☐	☐
(2) Pre-designation of a properly qualified oil discharge response coordinator who is charged with the responsibility and delegated commensurate authority for directing and coordinating response operations and who knows how to request assistance from Federal authorities operating under existing national and regional contingency plans.	☐	☐
(3) A preplanned location for an oil discharge response operations center and a reliable communications system for directing the coordinated overall response operations.	☐	☐
(4) Provisions for varying degrees of response effort depending on the severity of the oil discharge.	☐	☐
(5) Specification of the order of priority in which the various water uses are to be protected where more than one water use may be adversely affected as a result of an oil discharge and where response operations may not be adequate to protect all uses.	☐	☐
(e) Specific and well defined procedures to facilitate recovery of damages and enforcement measures as provided for by State and local statutes and ordinances.	☐	☐

[17] The contingency plan should be consistent with all applicable state and local plans, Area Contingency Plans, and the NCP.

ATTACHMENT D: TIER II QUALIFIED FACILITY CHECKLIST ☐ NA

TIER II QUALIFIED FACILITY PLAN REQUIREMENTS —40 CFR 112.6(b)		
112.6(b)(1)	**Plan Certification:** Owner/operator certified in the Plan that:	☐ Yes ☐ No
(i)	He or she is familiar with the requirements of 40 CFR part 112	☐ Yes ☐ No ☐ NA
(ii)	He or she has visited and examined the facility[18]	☐ Yes ☐ No ☐ NA
(iii)	The Plan has been prepared in accordance with accepted and sound industry practices and standards and with the requirements of this part	☐ Yes ☐ No ☐ NA
(iv)	Procedures for required inspections and testing have been established	☐ Yes ☐ No ☐ NA
(v)	He or she will fully implement the Plan	☐ Yes ☐ No ☐ NA
(vi)	The facility meets the qualification criteria set forth under §112.3(g)(2)	☐ Yes ☐ No ☐ NA
(vii)	The Plan does not deviate from any requirements as allowed by §§112.7(a)(2) and 112.7(d), except as described under §112.6(b)(3)(i) or (ii)	☐ Yes ☐ No ☐ NA
(viii)	The Plan and individual(s) responsible for implementing the Plan have the full approval of management and the facility owner or operator has committed the necessary resources to fully implement the Plan.	☐ Yes ☐ No ☐ NA
112.6(b)(2)	**Technical Amendments:** The owner/operator self-certified the Plan's technical amendments for a change in facility design, construction, operation, or maintenance that affected potential for a §112.1(b) discharge	☐ Yes ☐ No ☐ NA
If YES	• Certification of technical amendments is in accordance with the self-certification provisions of §112.6(b)(1).	☐ Yes ☐ No ☐ NA
(i)	A PE certified a portion of the Plan (i.e., Plan is informally referred to as a hybrid Plan)	☐ Yes ☐ No ☐ NA
If YES	• The PE also certified technical amendments that affect the PE certified portion of the Plan as required under §112.6(b)(4)(ii)	☐ Yes ☐ No ☐ NA
(ii)	The aggregate aboveground oil storage capacity increased to more than 10,000 U.S. gallons as a result of the change	☐ Yes ☐ No ☐ NA
If YES	*The facility no longer meets the Tier II qualifying criteria in §112.3(g)(2) because it exceeds 10,000 U.S. gallons in aggregate aboveground storage capacity.*	
	The owner/operator prepared and implemented a Plan within 6 months following the change and had it certified by a PE under §112.3(d)	☐ Yes ☐ No ☐ NA
112.6(b)(3)	**Plan Deviations:** Does the Plan include environmentally equivalent alternative methods or impracticability determinations for secondary containment?	☐ Yes ☐ No ☐ NA
If YES	Identify the alternatives in the hybrid Plan:	
	• Environmental equivalent alternative method(s) allowed under §112.7(a)(2);	☐ Yes ☐ No ☐ NA
	• Impracticability determination under §112.7(d)	☐ Yes ☐ No ☐ NA
112.6(b)(4)	• For each environmentally equivalent measure, the Plan is accompanied by a written statement by the PE that describes: the reason for nonconformance, the alternative measure, and how it offers equivalent environmental protection in accordance with §112.7(a)(2);	☐ Yes ☐ No ☐ NA
	• For each secondary containment impracticability determination, the Plan explains the reason for the impracticability determination and provides the alternative measures to secondary containment required in §112.7(d)	☐ Yes ☐ No ☐ NA
	AND	
(i)	PE certifies in the Plan that:	
(A)	He/she is familiar with the requirements of 40 CFR Part 112	☐ Yes ☐ No ☐ NA
(B)	He/she or a representative agent has visited and examined the facility	☐ Yes ☐ No ☐ NA
(C)	The alternative method of environmental equivalence in accordance with §112.7(a)(2) or the determination of impracticability and alternative measures in accordance with §112.7(d) is consistent with good engineering practice, including consideration of applicable industry standards, and with the requirements of 40 CFR Part 112.	☐ Yes ☐ No ☐ NA

Comments:

---

[18] Note that only the person certifying the Plan can make the site visit

Appendix G
Toolkit for Evaluating FIFRA

This appendix provides a series of checklists intended to assist compliance managers and personnel performing inspections and assessments of some of the more common operations involving selling, producing, importing and exporting pesticides as governed by FIFRA. These tools are provided for instructional purpose and guidance only and are not intended to represent an all-inclusive collection of FIFRA regulatory requirements. The tools provided in this appendix are adapted from EPA and organized within this appendix as follows:

FIFRA:

1. Market Place Inspection
2. Producer Inspection
3. Pesticide Container Inspection
4. Restricted Use Pesticide Checklist

TABLE G.1
FIFRA Market Place Checklist

Date: _____ Time: _____

Inspector's name: _____ Signature: _____

Pesticide Label, Device Label, Labeling and Packaging Inspection

1. Pesticides packaged, labeled and released for shipment/FIFRA Section 9(a)
   a. Brand name(s): _____
   b. EPA registration number(s): _____
   c. Label and labeling review/40 CFR 156: _____
2. Child-resistant packaging requirements for each pesticide product packaged, labeled and distributed-sold/40 CFR 157
   a. Criteria requiring child-resistant packaging: _____
   b. Unit packaging: _____
3. Records of shipment of pesticides, devices and active ingredients/40 CFR 169.3
   a. Brand name of pesticide or device: _____
   b. Name of originating carrier: _____
   c. Date shipped or delivered for shipment:_____
   d. Quantities shipped or delivered for shipment:_____
   e. Bill of lading, shipping records, invoices, etc.:_____
4. Other records
   a. Copies of all domestic advertising of pesticide/40 CFR 168.22:_____
   b. Copies of guarantees given pursuant to FIFRA Section 12(b)(1):_____

TABLE G.2
FIFRA Producer Checklist

Date: _____ Time: _____
Inspector's name: _____ Signature: _____

Pesticide/Device Label, Labeling and Packaging Inspection
1. Pesticides packaged, labeled and released for shipment/FIFRA Section 9(a)?
   a. Brand name(s): _____
   b. EPA registration number(s): _____
   c. Label and labeling review/40 CFR 156: _____
2. Child-resistant packaging requirements for each pesticide product packaged, labeled and released for shipment/40 CFR 157
   a. Criteria requiring child-resistant packaging: _____
   b. Unit packaging: _____

Records Retention Requirements for Pesticide Production
3. Records of production of pesticides/40 CFR 169.2(a)
   a. Brand name: _____
   b. EPA registration number/experimental-use permit number: _____
   c. Amounts produced per batch:* _____
   d. Batch identification: _____
   e. Length of retention of records: _____
4. Records of production of devices/40 CFR 169.2(b)
   a. Brand name: _____
   b. Quantities of device(s) produced:* _____
5. Records of receipt of pesticides, devices and active ingredients/40 CFR 169.2(c):
   a. Brand name of pesticide, device or common or chemical name of active ingredient: _____
   _____
   b. Name and address of shipper: _____
   c. Name of delivering carrier: _____
   d. Date received by establishment: _____
   e. Quantities received: _____
   f. Retention of records: _____
6. Records of shipment of pesticides and active ingredient/40 CFR 169.2(d):
   a. Brand name of pesticide, or common or chemical name of active ingredient: _____
   _____
   b. Name and address of consignee: _____
   c. Where the pesticide under a FIFRA Section 5 Permit, Section 18 exemption or Section 24(c) registration is/was produced: _____
   d. Name of originating carrier: _____
   e. Date shipped or delivered for shipment: _____
   f. Quantities shipped or delivered for shipment: _____
   g. Retention of records: _____
7. Other records/40 CFR 169.2
   a. Inventory records: _____
   b. Copies of all domestic advertising of the restricted uses of any pesticide registered for restricted use: _____
   c. Retention of records: _____
   d. Copies of guarantees given pursuant to FIFRA Section 12(b)(1): _____
   e. Retention of records: _____
   f. Records on the method of disposal, dates of disposal, location of the disposal site(s) and the types and amounts of pesticides, device or active ingredients disposed of by the producer/contractor: _____
   g. FIFRA records: _____
   h. Retention of records: _____
   i. RCRA records: _____

(Continued)

TABLE G.2 (CONTINUED)
FIFRA Producer Checklist

8. Records retention requirements of child-resistant packaging/40 CFR 157.36
   a. Description of the packaging for each registration: _____
   b. Certification statement for each registration: _____
   c. Test data verification of child resistance for each registration: _____
   d. Records verifying that the packaging meets the compatibility and durability standards:_____

9. Records requirements under the pesticide export policy/40 CFR 168, Subpart D:
   a. Confidential statement of formula for pesticides not registered under FIFRA:* _____
   b. Specifications or directions of the foreign purchaser: _____
   c. Bilingual labeling and other required labeling:_____
   d. Foreign purchaser acknowledgment statements:_____
   e. Establishment/foreign purchaser contracts: _____

Records Reporting Requirements for Registrants/Pesticide-Producing Establishments
10. Application for amended registration/40 CFR 152.44: _____
11. Modifications to registration not requiring amended applications/40 CFR 152.46: _____
_____
12. Currency of address of record and authorized agent/40 CFR 152.122: _____
_____
13. Submission of information pertaining to adverse effects/FIFRA Section 6(a)(2), 40 CFR 152.125 and 153, Subpart D: _____
14. Supplemental distribution/40 CFR 152.132: _____
15. Transfer of registration(s)/40 CFR 152.135: _____
16. Initial production reports/Section 7(c) and 40 CFR 167:* _____
17. Annual production reports/Section 7(c) and 40 CFR 167:*_____
18. Notification of stored pesticides with canceled or suspended registrations/FIFRA Section 6(g):
_____

Policy on Bulk Pesticide Repackaging
19. Bulk repackaging/contract bulk repackaging agreements: _____
20. Complete end-use labeling/bulk containers: _____
21. Complete end-use labeling/mini-bulk containers: _____

Additional Information
22. Private labeling by a contractor: _____
23. Private labeling under contract for another registrant/distributor: _____
24. Contract manufacturing agreements: _____

Technical or Manufacturing Use Pesticides or Chemicals
25. Do all products that require registration have the appropriate EPA registration number, EPA establishment number and proper labeling?   Yes___ No___
26. Are invoices, shipping records, bills of lading manifests and common carrier records that pertain to each technical or manufacturing use product available?   Yes___ No___
27. Are any contract manufacturing agreements where the establishment is producing a pesticide from another firm, etc. (e.g., supplemental distributor, supplementally registered product, distributor product) maintained and available?   Yes___ No___

TABLE G.3
FIFRA Container Checklist

Date: _____ Time: _____
Inspector's name: _____ Signature: _____

RECORDS FOR REFILLABLE CONTAINER PACKAGING
Record-Keeping (Registrants Who Distribute in Bulk to Independent Refillers), 40 CFR 165(i) (1), 165 (h)(2) and (3)

   1. Are the following records being kept (for each product that the facility is refilling) for the current operating year and maintained for three years?   Yes___ No___
      a. Residue removal procedure for each product distributed in bulk?
      b. A list of acceptable containers for each product distributed in bulk?
      c. A written contract with every independent refiller that is repackaging any of these products?

Record-Keeping for Refillers (Both Independent Refillers and Registrants Who Sell or Distribute Directly in Refillable Containers, 40 CFR 165.65(d)(10) and (i)(1), 165.70(e)(10))

   2. Has the refiller kept the following records (for each product the facility is refilling) for the current operating year and maintaining them for three years?   Yes___ No___
      a. Residual removal procedure?
      b. List of acceptable containers?

Additional Requirements for Independent Refillers Who Are Not Registrants, 40 CFR 165.70(b), (e)(5)(i), (e)(13),(e)(14) and (j)(2)

   3. Does the refiller have a written contract from the pesticide's registrant to repackage the registrant's pesticide?   Yes___ No___
   4. Did the refiller keep records of the written contract from the registrant for the previous three years?   Yes___ No___

Record-Keeping for Refillers (Both Independent and Registrants Who Sell or Distribute Directly in Bulk) Each Time Pesticide Is Repackaged into Refillable Containers, 40 CFR 165.70 (e)(10) and (j)(2)

   5. Does the refiller keep a record of the following, each time a pesticide is repackaged?
                                                                                                      Yes___ No___
      a. EPA registration number of pesticide
      b. The date of repackaging
      c. Serial number or other identifying code of the container
   6. Does the refiller maintain records for three years after the date of packaging?   Yes___ No___

Other Requirements for Refillers (Both Independent Refillers and Registrants Who Sell or Distribute Directly in Bulk) 40 CFR 165(d)(11) and (12), 165.70(e)(11) and (12)

   7. Does the refiller maintain records as required by 40 CFR Part 169 (Books and Records of Pesticide Production and Distribution)?   Yes___ No___
   8. Does the refiller report as required by 40 CFR 167 (Registration of Pesticide and Active Ingredient Producing Establishments: Submission of Pesticide Reports)?   Yes___ No___

RECORD-KEEPING FOR CONTAINER DESIGN: NONREFILLABLES
Record-Keeping 40 CFR 165.27(b)

   9. Does the registrant have records to show compliance with the requirements for:
      a. Standard closure?   Yes___ No___
      b. Container dispensing?   Yes___ No___
      c. Residue removal?   Yes___ No___

TABLE G.4
FIFRA Restricted Use Pesticide Applicator Checklist

Date: _____Time: _____

Inspector's name: _____ Signature: _____

For commercial applicators and those who contract with commercial applicators to apply RUPs to property owned by another person.

1. Are the following applicator records being maintained for at least 24 months from the date of pesticide use? 40 CFR 171.11(c)(7) Yes___ No___

   a. Name and address of the person for whom the pesticide was applied?

   b. Name and address of the person for whom the pesticide was applied?

   c. Target pests?

   d. Specific crop or commodity and site to which the pesticide was applied

   e. Date and time of application

   f. Product/trade name and EPA Registration Number of the pesticide applied

   g. Amount of pesticide applied and percentage of active ingredient per unit of pesticide used

   h. Disposal information

Appendix H
Toolkit for Evaluating TSCA

This appendix provides a series of checklists intended to assist compliance managers and personnel performing certain inspections and assessments of some of the typical operations involving asbestos and PCB inspections governed by TSCA. These tools are provided for instructional purpose and guidance only and are not intended to represent an all-inclusive collection of TSCA regulatory requirements. The tools provided in this appendix are adapted from EPA checklists and guidance. The example checklists provided in this appendix are as follows.

TSCA Compliance Inspections:

1. Asbestos Management
2. PCB Storage

TABLE H.1
Asbestos Management Plan Checklist

Date: _____ Time: _____
Designated person: _____ Signature: _____
General

1. Has an Asbestos Management Plan been developed for your school? (40 CFR 763.93) — Yes___ No___
2. Does the Local Education Agency (LEA) have a complete and up-to-date copy of the school's management plan in both the LEA's administrative office and the school's administrative office? (40 CFR 763.93(g)(2)-(3)) — Yes___ No___
3. Was the management plan developed by an accredited management planner? (40 CFR 763.93(e)) — Yes___ No___
4. For each consultant who contributed to the management plan, does the plan include the following: (40 CFR 763.93 (e)(12)(i)-(ii)) — Yes___ No___
   a. Consultant's name?
   b. A statement that he/she is accredited under the state accreditation program or another state's accreditation program or an EPA-approved course? — Yes___ No___
5. Does the management plan include a list of the name and address of each building used as a school building and identify whether the school building has the following: (40 CFR 763.93(a)(1)-(2) and 763.93(e)(1)) — Yes___ No___
   a. Friable ACBM (asbestos-containing building material)?
   b. Nonfriable ACBM?
   c. Friable and nonfriable suspected ACBM assumed to be ACM (asbestos-containing material)?
6. If a new school building was constructed after October 12, 1988, and is asbestos-free, does the management plan include the following and has a copy of same been provided by the LEA to the EPA Regional Office: (40 CFR 763.99(a)(7)) — Yes___ No___
   a. A statement signed by an architect or project engineer responsible for the construction of the building, or by an accredited inspector, indicating that no ACBM was specified as a building material in any construction document for the building or, to the best of his/her knowledge, no ACBM was used as a building material in the building?
7. Does the management plan include a copy of any of the statements required under 40 CFR 763.99(a)(1)-(7) to support an exclusion from inspection that the school may qualify for under 40 CFR 763.99 and has a copy of any such statement been provided by the LEA to the Regional Office? (40 CFR 763.99) — Yes___ No___
8. Does the management plan include the following information about the LEA Designated Person (DP): (40 CFR 763.93(e)(4) and (i)) — Yes___ No___
   a. Name, address and telephone number of the DP?
   b. Course name, dates and hours of training that the DP attended to carry out his/her AHERA duties?
   c. Signed statement by the DP that the LEA's general responsibilities under 40 CFR 763.84 have been or will be met?
9. Does the management plan include the following recommendations: (40 CFR 763.93(e)(9)) — Yes___ No___
   a. A plan for reinspection required under 40 CFR 763.85?
   b. A plan for operations and maintenance activities (including initial cleaning) required under 40 CFR 763.91?
   c. A plan for periodic surveillance required under 40 CFR 763.92?
   d. A description of the management planner's recommendation for additional cleaning under 40 CFR 763.91(c)(2), as part of an operations and maintenance program, and the response of the LEA to that recommendation?
10. Does the management plan include an evaluation of resources needed to carry out response actions, reinspections, operations and maintenance and periodic surveillance and training? (40 CFR 763.93(e)(11)) — Yes___ No___

(Continued)

TABLE H.1 (CONTINUED)
Asbestos Management Plan Checklist

11. Does the management plan include a record of the minimum two hours of
    awareness training required under 40 CFR 763.92(a)(1) for all maintenance and
    custodial staff who may work in a building that contains ACBM, whether or not
    they are required to work with ACBM and does the record include the following
    information: (40 CFR 763.93(h) and 763.94(c))   Yes___ No___
    a. Person's name and job title?
    b. Date training was completed?
    c. Location of training?
    d. Number of hours completed?
12. Does the management plan include a record of the additional 14 hours of training
    required under 40 CFR 763.92(a)(2) for maintenance and custodial staff who conduct
    any activities that will result in the disturbance of ACBM and does the record include
    the following information: (40 CFR 763.93(h) and 763.94(c))   Yes___ No___
    a. Person's name and job title?
    b. Date training was completed?
    c. Location of training?
    d. Number of hours completed?

Inspections and Reinspections

13. For inspections conducted before December 14, 1987 (i.e., the effective date of the
    October 30, 1987, EPA Asbestos-Containing Materials in Schools rule), does the
    management plan include the following information: (40 CFR 763.93(e)(2)(i)-(v))
                                                                               Yes___ No___
    a. Date of inspection?
    b. Blueprint, diagram or written description of each school building that identifies
       clearly each location and approximate square or linear footage of
       homogeneous/sampling area sampled for ACM?
    c. If possible, the exact locations where the bulk samples were collected and the
       dates of collection?
    d. A copy of the analyses of any bulk samples, dates of analyses and a copy of any
       other laboratory reports pertaining to the analyses?
    e. Description of response actions or preventive measures taken, including, if
       possible, the names and addresses of all contractors, start and completion dates
       and air clearance sample results?
    f. Description of assessments of material identified prior to December 14, 1987,
       as friable ACBM or friable suspected ACBM assumed to be ACM and the
       name, signature, state of accreditation and if, applicable, the accreditation
       number of the person making the assessments (i.e., inspector)?
14. Does the management plan include for each inspection and reinspection conducted
    under 40 CFR 763.85 the following information: (40 CFR 763.93(e)(3)(i))   Yes___ No___
    a. Date of the inspection or reinspection?
    b. Name, signature, state of accreditation and, if applicable, the accreditation
       number for each accredited inspector performing the inspection or reinspection?
15. Does the management plan include for each inspection and reinspection conducted
    under 40 CFR 763.85 the following sampling information: (40 CFR 763.93(e)(3)
    (ii)-(iii))   Yes___ No___
    a. Blueprint, diagram or written description of each school building that identifies
       clearly each location and approximate square or linear footage of homogeneous
       areas where material was sampled for ACM?
    b. Exact location where each bulk sample was collected and the date of collection
       of each bulk sample?
    c. Homogeneous areas where friable suspected ACBM is assumed to be ACM?
    d. Homogeneous areas where nonfriable suspected ACBM is assumed to be ACM?
    e. Description of the manner used to determine sampling locations?
    f. The name, signature, state of accreditation and, if applicable, the accreditation
       number for each accredited inspector that collected samples?

(Continued)

TABLE H.1 (CONTINUED)
Asbestos Management Plan Checklist

16. Does the management plan include for each inspection and reinspection conducted under 40 CFR 763.85 the following information on the analysis of the bulk samples and has it been submitted to the DP for inclusion in the plan within 30 days of the analysis: (40 CFR 763.87(d) and 763.93(e)(3)(iv))   Yes___ No___
    a. Copy of the analysis of any bulk samples collected and analyzed?
    b. Name and address of any laboratory that analyzed bulk samples?
    c. A statement that any laboratory used meets the applicable laboratory accreditation requirements of 40 CFR 763.87(a)?
    d. Dates of any analyses performed?
    e. Name and signature of the person performing each analysis?
17. Does the management plan include for each inspection and reinspection conducted under 40 CFR 763.85 the following assessment information and has it been submitted to the DP for inclusion in the plan within 30 days of the assessment: (40 CFR 763.88(a)(2) and 763.93(e)(3)(v)   Yes___ No___
    a. Written assessments (signed and dated) required to be made under 40 CFR 763.88 of all ACBM and suspected ACBM assumed to be ACBM?
    b. Name, signature, state of accreditation and, if applicable, the accreditation number of each accredited person making the assessment (i.e., inspector(s))
18. Has the following information about the inspection been recorded and submitted to the DP for inclusion in the management plan within 30 days of the inspection: (40 CFR 763.85(a)(4)(vi)(A)-(E) and 763.88(a)(2))   Yes___ No___
    a. Inspection report with the date of inspection signed by each accredited inspector making the inspection, the state of accreditation and if applicable, his/her accreditation number?
    b. Inventory of the locations of the homogeneous areas where samples are collected, exact location where each bulk sample is collected, dates that samples are collected, homogeneous areas where friable suspected ACBM is assumed to be ACM and homogeneous areas where nonfriable suspected ACBM is assumed to be ACM?
    c. Description of the manner used to determine sampling locations, the name and signature of each accredited inspector who collected the samples, state of accreditation and, if applicable, his/her accreditation number?
    d. List of whether the homogeneous areas identified under 40 CFR 763.85(a)(4) (vi)(B) of this section are surfacing material, thermal system insulation or miscellaneous material?
    e. Assessments of friable material (signed and dated), the name and signature of each accredited inspector making the assessment, state of accreditation and, if applicable, his/her accreditation number?
19. Has the following information about the reinspection been recorded and submitted to the DP for inclusion in the management plan within 30 days of the reinspection: (40 CFR 763.85(b)(3)(vii)(A)-(C) and 763.88(a)(2))   Yes___ No___
    a. Date of reinspection, name and signature of the person making the reinspection, state of accreditation and, if applicable, his/her accreditation number and any changes in the condition of known or assumed ACBM?
    b. Exact location where samples were collected during the reinspection, a description of the manner used to determine sampling locations, the name and signature of each accredited inspector who collected the samples, state of accreditation and, if applicable, his/her accreditation number?
    c. Any assessments or reassessments of friable material, date of the assessment or reassessment, the name and the signature of the accredited inspector making the assessments, state of accreditation and, if applicable, his/her accreditation number?

(Continued)

TABLE H.1 (CONTINUED)
Asbestos Management Plan Checklist

Response Actions

20. Does the management plan include the recommendations made to the LEA regarding response actions under 40 CFR 763.88(d) and the following information about the accredited management planner: (40 CFR 763.88(d) and 763.93(e)(5)) Yes___ No___

    a. Name, signature, state of accreditation and, if applicable, the accreditation number for each accredited management planner making the recommendations?

21. Does the management plan include a detailed description of preventive measures and response actions to be taken, including the following: (40 CFR 763.93(e)(6)) Yes___ No___

    a. Methods to be used for any friable ACBM?
    b. Locations where such measures and actions will be taken?
    c. Reasons for selecting the response action or preventive measure?
    d. Schedule for beginning and completing each preventive measure or response action?

22. Does the management plan include one of the following statements for the person or persons who inspected for ACBM and who will design or carry out response actions, except for operations and maintenance, with respect to the ACBM: (40 CFR 763.93(e)(7))   Yes___ No___

    a. Statement that he/she is accredited under the state accreditation program, or that the LEA has used (or will use) persons accredited under another state's accreditation program or an EPA-approved course?

23. Does the management plan include a detailed written description of each preventive measure and response action taken for friable and nonfriable ACBM and friable and nonfriable suspected ACBM assumed to be ACM, including the following: (40 CFR 763.94(b)(1))   Yes___ No___

    a. Methods used?
    b. Location where the measure or action was taken?
    c. Reasons for selecting the measure or action?
    d. Start and completion dates of the work?
    e. Names and addresses of all contractors involved and, if applicable, their state of accreditation and accreditation numbers?
    f. If ACBM is removed, the name and location of storage or disposal site of the ACM?

24. Does the management plan include the following sampling information required to be collected at the completion of certain response actions specified by 40 CFR 763.90(i): (40 CFR 763.94(b)(2))   Yes___ No___

    a. Name and signature of any person collecting any air sample required to be collected?
    b. Locations where samples were collected?
    c. Date of collection?
    d. Name and address of the laboratory analyzing the samples?
    e. Date of analysis?
    f. Results of analysis?
    g. Method of analysis?
    h. Name and signature of the person performing the analysis?
    i. Statement that the laboratory meets the applicable laboratory accreditation requirements of 40 CFR 763.90(i)(2)(ii)?

25. Does the management plan include a detailed description in the form of a blueprint, diagram or written description of any ACBM or suspected ACBM assumed to be ACM that remains in the school once response actions are undertaken under 40 CFR 763.90 and is the description updated as response actions are completed? (40 CFR 763.93(e)(8))   Yes___ No___

(Continued)

TABLE H.1 (CONTINUED)
Asbestos Management Plan Checklist

26. For each homogeneous area where all ACBM have been removed, have records been retained in the management plan for at least three years after the next reinspection required under 40 CFR 763.85(b)(1) or for an equivalent period? (40 CFR 763.93(h) and 763.94(a))   Yes___ No___

Operations and Maintenance

27. Does the management plan include a record of each cleaning conducted under 40 CFR 763.91(c), including the following: (40 CFR 763.93(h) and 763.94(e))
    Yes___ No___
    a. Name of each person performing the cleaning?
    b. Date of the cleaning?
    c. Locations cleaned?
    d. Methods used to perform the cleaning?

28. Does the management plan include a record of each O&M activity and major asbestos activity, with the following information: (40 CFR 763.93(h) and 763.94(f) and(g))   Yes___ No___
    a. Name of each person performing the activity?
    b. For a major asbestos activity, the name, signature, state of accreditation and, if applicable, the accreditation number of each person performing the activity?
    c. Start and completion date of each activity?
    d. Location of the activity?
    e. Description of the activity including preventive measures used?
    f. If ACBM is removed, the name and location of the storage and disposal site for the ACM?

29. Does the management plan include a record of each fiber release episode, whether major or minor, with the following information: (40 CFR 763.93(h) and 763.94(h))   Yes___ No___
    a. Date and location of the episode?
    b. Method of repair?
    c. Preventive measure or response action taken?
    d. Name of each person performing the work?
    e. If ACBM is removed, the name and location of the storage and disposal site of the ACM?

Periodic Surveillance

30. Does the management plan include a record of each periodic surveillance performed under 40 CFR 763.92(b), with the following information: (40 CFR 763.92(b)(2)(ii)-(iii), 763.93(h) and 763.94(d))   Yes___ No___
    a. Name of person performing the surveillance?
    b. Date of the surveillance?
    c. Any changes in the condition of the material?

31. Does the management plan include the following notification information: (40 CFR 763.93(e)(10) and 763.93(g)(4))   Yes___ No___
    a. Description of the steps taken to notify, in writing, at least once a year, parent, teacher and employee organizations of the availability of the management plan for review?
    b. Dated copies of all such management plan availability notifications (e.g., letter, newsletter)?
    c. Description of the steps taken to inform workers and building occupants or their legal guardians about inspections, reinspections, response actions and post-response action activities, including periodic reinspection and surveillance activities that are planned or are in progress? (Under 40 CFR 763.84(c), the LEA must inform them about these activities at least once each school year.)

TABLE H.2
PCB Storage Checklist

Date: _____ Time: _____

Designated person: _____ Signature: _____

1. Was the last inspection conducted within the previous 30 days?	Yes___ No___
2. Date of last inspection	
3. Are all containers marked with the large PCB mark?	Yes___ No___
4. Number of containers in storage area?	
5. Are all entrances to the storage area marked with the large PCB mark?	Yes___ No___
6. Are nonradioactive PCBs in this area stored in containment with a continuous curb that is at least 6 inches in height, or are the containers stored within a salvage container?	Yes___ No___
7. Does the containment have a volume of at least two times the internal volume of the largest PCB article or PCB container, or 25% of the total volume of all PCB articles or PCB containers stored, whichever is greatest?	Yes___ No___
8. Have salvage containers been opened and inspected for leaks during this quarter (Le. Jan–Mar; Apr–Jun; Jul–Sep; Oct–Dec)?	Yes___ No___
9. Is the floor free from any openings such as drain valves, floor drains, expansion joints, lines, etc.?	Yes___ No___
10. Are the floor and containment curbing constructed of continuous smooth and impervious materials, such as Portland cement concrete or steel?	Yes___ No___
11. Are roof and walls adequate to prevent rainwater from reaching the stored PCBs and PCB items?	Yes___ No___
12. Are all containers in good condition?	Yes___ No___
13. Is the 'Removal from Service for Disposal' date recorded on the documentation for each PCB item in each container?	Yes___ No___
14. Are all PCBs being stored for less than nine months?	Yes___ No___

APPENDIX I
Toolkit for Evaluating NEPA

This appendix provides a series of checklists intended to assist compliance managers and personnel performing inspections and assessments of some of the more common operations involving categorical exclusions, environmental assessments and environmental impact statements as governed by NEPA. These tools are provided for instructional purpose and guidance only and are not intended to represent an all-inclusive collection of NEPA regulatory requirements. It is noted that federal agencies establish agency-specific procedures and guidelines for evaluating projects under NEPA. The tools provided in this appendix are adapted from EPA, Department of Energy and the Federal Transit Administration and organized within this appendix as follows.

NEPA:

1. NEPA Compliance Checklist
2. Categorical Exclusion Checklist
3. Environmental Assessment Checklist
4. Environmental Impact Statement Checklist

TABLE I.1
NEPA Compliance Checklist

Date: _____ Time: _____

Inspector's name: _____ Signature: _____

1. Have significant adverse effects on public health or safety?   Yes___ No___

2. Have significant adverse effects on natural resources and unique geographic?   Yes___ No___

3. Characteristics as historic or cultural resources; park, recreation or refuge lands; wilderness areas; wild or scenic rivers; national natural landmarks; sole or principal drinking water aquifers; prime farmlands; wetlands (Executive Order 11990); floodplains (Executive Order 11988); national monuments; migratory birds (Executive Order 13186); and other ecologically significant or critical areas under federal ownership or jurisdiction?   Yes___ No___

4. Have highly controversial environmental effects or involve unresolved conflicts concerning alternative uses of available resources [NEPA Section 102(2)(E)]?   Yes___ No___

5. Have highly uncertain and potentially significant environmental effects or involve unique or unknown environmental risks?   Yes___ No___

6. Have a precedent for future action or represent a decision in principle about future actions with potentially significant environmental effects?   Yes___ No___

7. Have a direct relationship to other actions with individually insignificant but cumulatively significant environmental effects?   Yes___ No___

8. Have significant adverse effects on properties listed or eligible for listing on the National Register of Historic Places as determined by either the bureau or office, the State Historic Preservation Officer, the Tribal Historic Preservation Officer, the Advisory Council on Historic Preservation or a consulting party under 36 CFR 800?   Yes___ No___

9. Have significant adverse effects on species listed, or proposed to be listed, on the list of Endangered or Threatened Species, or have significant adverse effects on designated critical habitat for these species?   Yes___ No___

10. Have the possibility of violating a federal law, or a state, local or tribal law or requirement imposed for the protection of the environment?   Yes___ No___

11. Have the possibility for a disproportionately high and adverse effect on low-income or minority populations (Executive Order 12898)?   Yes___ No___

12. Have the possibility to limit access to and ceremonial use of Indian sacred sites on federal lands by Indian religious practitioners or significantly adversely affect the physical integrity of such sacred sites (Executive Order 13007)?   Yes___ No___

13. Have the possibility to significantly contribute to the introduction, continued existence or spread of noxious weeds or nonnative invasive species known to occur in the area or actions that may promote the introduction, growth or expansion of the range of such species (Federal Noxious Weed Control Act and Executive Order 13112)?   Yes___ No___

If any of the above extraordinary circumstances receive a 'Yes' answer, an EA must be prepared.

TABLE I.2
NEPA Categorical Exclusion Checklist

Date: _____ Time: _____

Inspector's name: _____ Signature: _____

Information Required for Probable Categorical Exclusion

(23 CFR Part 771.118)

A. **Detailed Project Description:** Describe the project including the type (such as bus storage, maintenance and/or administration facilities). Indicate the size of the proposed facility, number of vehicles and staff it will house. Describe any construction, demolition and soil excavation activities. Include a brief discussion summarizing the purpose and need for the project (e.g., congestion, state of good repair). Explain in common language how implementation of the project will address the project need and its proposed use. Include a complete description of the project components such as length of the project in feet or miles, property size, history, ownership information (land management authority) and acreage, and document previously conducted studies if applicable. Provide graphics that describe the proposed project.

B. **Location (Including Address):** Attach a project location map or diagram, such as a USGS topographic map that identifies the project location. Clearly delineate the project and include streets and features specifically called out in the 'detailed project description.' If the project work occurs at more than one location, include those locations and adjoining parcels on the map. This information is partly used to determine the probability of impact on the human and natural environment.

C. **Metropolitan Planning and Air Quality Conformity:** Is the proposed project included in the current planning documents, either exclusively or in a grouping of projects or activities? What is the conformity status of that plan? Is the proposed project, or appropriate phases of the project, included in the implementation plan (IP)? What is the conformity status of the IP? Is the project located in an air quality nonattainment area? Is the project exempt from a conformity review per Table 2 of 40 CFR 93.126? Refer to the nonattainment/maintenance area maps at the U.S. EPA website to determine if the project is located in an area that meets all National Ambient Air Quality Standards.

D. **Land Use and Zoning:** Describe property zoning and consistency with proposed use. Attach a zoning map of the project area and surrounding area. Attach a land use map that identifies land and water uses in the project area. This information is partly used to determine the probability of impact on the human and natural environment. Land use plans and zoning maps can be obtained from the tax assessor, city, county or metropolitan planning organizations.

E. **Traffic Impacts:** Describe potential traffic impacts, including short-term impacts during construction or demolition, and whether the existing roadways have adequate capacity for increased bus and other vehicular traffic as part of the proposed project. Examples of construction-related impacts include lane closures, detours or dust abatement requirements. Briefly describe traffic control measures required to minimize impacts of construction.

F. **CO Hot Spots:** If there are serious traffic impacts at any affected intersection or area where buses congregate, and if the area is in an air quality nonattainment area for CO, demonstrate that CO hot spots will not be created as a result of the project.

(Continued)

TABLE I.2 (CONTINUED)
NEPA Categorical Exclusion Checklist

G. **PM2.5 and PM10 Hot Spots:** If there are serious traffic impacts at any affected intersection or area where buses congregate, and if the area is a nonattainment or maintenance area for any particulate matter (PM2.5 or PM10), then demonstrate that PM2.5 or PM10 'hot spots' will not result. In nonattainment areas, interagency concurrence and documentation must be attached. If the proposed project is not in a nonattainment or maintenance area for PM2.5 and PM10, then state this in the discussion. Refer to the nonattainment/maintenance area maps at the U.S. EPA website to determine if the project is located in an area that meets all National Ambient Air Quality Standards.

H. **Historic Resources:** Describe any cultural, historic or archaeological resources located in the immediate vicinity of the proposed project and the impact of the project on the resources. Show these resources on a map. All consultations are initiated per Section 106 of the National Historic Preservation Act (NHPA). Agency also makes a determination of 'No Effect/No Historic Properties' or 'No Historic Properties Affected,' if no historic resources or potential to affect resources exists. Agency requests concurrence for this determination from the appropriate State Historic Preservation Office (SHPO) or Tribal Historic Preservation Office (THPO). SHPO/THPO concurrence *must* be included as an attachment before NEPA approval. If an 'Adverse Effect' determination is made as a result of the proposed project, rather than a 'No Effect/No Historic Properties' or 'No Historic Properties Affected' determination, then FTA may determine a new NEPA class of action to evaluate alternatives or mitigation measures to deter these adverse effects. If the project has potential effects to NRHP-eligible or listed projects, the Section 106 process must be followed. Refer to the ACHP website for more information. Projects involving modifications to historic buildings or structures should comply with the *Secretary of the Interior Standards for the Rehabilitation of Historic Structures*, which is available from the SHPO/THPO and the National Park Service.

I. **Visual Quality:** Describe the existing visual setting, identify any sensitive views/viewers, and describe the visual impact of the proposed project.

J. **Noise:** Compare distance between the center of the proposed project and the nearest noise receptor to the screening distance for this type of project. If the screening distance is not achieved, attach a 'General Noise Assessment' with conclusions.

K. **Vibration:** If the proposed project includes new or relocated steel rails/tracks, compare the distance between the center of the proposed project and the nearest vibration receptor to the screening distance for this type of project in agency's guidelines. If the screening distance is not achieved, attach a 'General Vibration Assessment' with conclusions.

L. **Acquisitions and Relocations Required:** Describe land acquisitions and displacements of residences and businesses. Include current use, ownership and the date and type of property transaction (such as lease or purchase). If Agency funds are used to acquire property or the property is used as local match, then the Uniform Relocation Assistance and Real Property Acquisition Policies Act of 1970 must be followed and documented. No offers or appraisals may occur prior to Agency's approval of a NEPA evaluation.

M. **Hazardous Materials:** If real property has been acquired, has a Phase I site assessment for contaminated soil and groundwater been performed? If a Phase II site assessment is recommended, has it been completed? What steps will be taken to ensure that human and ecological receptors in the project area are protected from contamination encountered during construction and operation of the project? State the results of consultation with the State agency with jurisdiction over proposed remediation of soil and/or groundwater contamination. Include anticipated effects of the project on asbestos-containing building materials and lead-based paints.

(Continued)

TABLE I.2 (CONTINUED)
NEPA Categorical Exclusion Checklist

N. **Social Impacts and Community Disruption:** Provide a socioeconomic profile of the affected community. Describe the impacts of the proposed project on the community. Identify any community resources that would be affected and the nature and extent of the effect.

O. **Environmental Justice:** Identify the concentrations of minority and low-income populations in the area. Following Agency guidelines on environmental justice, define 'minority' and 'low-income' populations, and describe whether or not the project would result in disproportionately high and adverse impacts on minority or low-income populations.

P. **Use of Public Parkland and Recreation Areas:** Indicate parks, recreational areas, wildlife refuges and/or trails on a project location map (Section 4(f) resources). Describe how the activities and purposes of these resources will be affected by the project. Based on the definitions of use outlined in 23 CFR § 774, determine if the project will result in an actual (direct), temporary or constructive (proximity impacts) use of the Section 4(f) resource. Locate Section 4(f) properties on project map.

Q. **Impacts on Wetlands:** Show potential wetlands and boundaries on a map. Integrate data from the National Wetlands Inventory. Describe the project's impact on on-site and adjacent wetlands. If the project impacts wetlands, provide documentation of consultations and permits from the U.S. Army Corps of Engineers, as well as minimization and mitigation efforts. If applicable, provide documentation to demonstrate that wetlands are not present, or the proposed project will not impact any wetland areas.

R. **Floodplain Impacts:** Determine if the project is within a 100-year floodplain. Review FEMA 100-year FIRMs on the FEMA website. Include a FIRM floodplain map, if available. Include all floodplain FIRM numbers that occur in the project area and the effective or revision date for each FIRM. Include the FEMA FIRM numbers for the project area, even if the 100-year floodplain has not been delineated. If the proposed project is located within the 100-year floodplain, describe what will be done to address possible flooding of the proposed project location and flooding induced by the project due to reduced capacity to retain storm water runoff. Provide documentation on how the project will be designed to restore floodplain capacity. If applicable, provide documentation to demonstrate that the project is not sited in a floodplain. If a determination cannot be made whether or not the project is within a 100-year floodplain, contact the county flood control district or the local floodplain manager for assistance.

S. **Impacts on Water Quality, Navigable Waterways and Coastal Zones:** If any of these resources are implicated, describe the project's potential impacts. Determine if National Pollutant Discharge Elimination System (NPDES) permits are applicable as a result of ground disturbance or point sources that will discharge pollutants into waters of the United States. Refer to BMPs at the U.S. EPA website. How will storm water be treated during and after construction? How will wastewater from bus washing facilities be treated? Determine if project area is in a sole-source aquifer; if not, document in narrative (refer to the U.S. EPA website).

T. **Impacts on Ecologically Sensitive Areas and Endangered Species:** Describe any natural areas (woodlands, prairies, wetlands, rivers, lakes, streams, designated wildlife or waterfowl refuges and geological formations) on or near the proposed project area. If present, state the results of consultation with the state department of natural resources and, if appropriate, the U.S. Fish and Wildlife Service on the impacts on critical habitats and on threatened and endangered fauna and flora that may be affected. Refer to the U.S. Fish and Wildlife Service website.

U. **Impacts on Safety and Security:** Describe the measures that would need to be taken to provide for the safe and secure operation of the project after its construction. List any security measures that are planned as part of the project (e.g., security guards, fencing, secured access, lighting, cameras, etc.)

V. **Impacts Caused by Construction:** Describe the construction plan and identify construction impacts with respect to noise, dust, utility disruption, debris and spoil disposal, air quality, water quality, erosion, safety and security and disruptions of traffic and access to businesses or residential property. Identify steps that will be taken to provide alternatives or mitigate the impacts of construction impacts. Cite applicable local, state and federal regulations and any standards or BMPs that will be followed. If applicable, please include any NPDES best practice measures (refer to the U.S. EPA website).

The action described above meets the criteria for a NEPA categorical exclusion (CE) in accordance with 23 CFR Part 771.118.

TABLE I.3
NEPA Environmental Assessment Checklist

List 1: General	Yes	No	N/A	PAGE	Adequacy Evaluation and Comments
1.1.0 SUMMARY					
1.1.1 Does the summary address the entire EA (Recommendations, p.3)?					
1.1.2 Is the summary consistent with information in the document?					
1.1.3 Does the summary highlight key differences among the alternatives?					
1.1.4 Does the summary describe:					
The underlying purpose and need for agency action?					
The proposed action?					
Each of the alternatives?					
The principal environmental issues and results?					
1.2.0 PURPOSE AND NEED FOR ACTION					
1.2.1 Does the statement of purpose and need define the need for Agency action (40 CFR 1508.9)					
1.2.2 Does the statement of purpose and need relate to the broad requirement or desire for agency action, and not to the need for one specific proposal?					
1.2.3 Is the statement of purpose and need written so that it does not inappropriately narrow the range of reasonable alternatives?					
1.2.4 Does the statement of purpose and need identify the problem or opportunity to which the agency is responding?					
1.3.0 DESCRIPTION OF THE PROPOSED ACTION AND ALTERNATIVES					
1.3.1 Is the proposed action described in sufficient detail so that potential impacts can be identified? Are all phases described (e.g., construction, operation, maintenance and decommissioning)?					
1.3.2 Are environmental releases associated with the proposed action quantified, including both the rates and durations?					
1.3.3 As appropriate, are mitigation measures included in the description of proposed action?					
1.3.4 Is the project description written broadly enough to encompass future modifications?					
1.3.5 Does the proposed action exclude elements that are more appropriate to the statement of purpose and need?					

(Continued)

TABLE I.3 (CONTINUED)
NEPA Environmental Assessment Checklist

List 1: General	Yes	No	N/A	PAGE	Adequacy Evaluation and Comments
1.3.6 Is the proposed action described in terms of the action to be taken (even a private action that has been federalized)?					
1.3.7 Does the EA address a range of reasonable alternatives that satisfy the agency's purpose and need, including reasonable alternatives outside Agency's jurisdiction?					
1.3.9 Does the EA include the no action alternative [10 CFR 1021.321(c)]?					
1.3.10 Is the no action alternative described in sufficient detail so that its scope is clear and potential impacts can be identified?					
1.3.11 Does the no action alternative include a discussion of the legal ramifications of no action, if appropriate?					
1.3.12 Does the EA take into account relationships between the proposed action and other actions to be taken by the agency in order to avoid improper segmentation?					
1.3.13 Does the proposed action comply with CEQ regulations for interim actions (40 CFR 1506.1)?					
1.4.0 DESCRIPTION OF THE AFFECTED ENVIRONMENT					
1.4.1 Does the EA *identify either the presence or absence* of the following within the area potentially affected by the proposed action and alternatives:					
Floodplains (EO 11988; 10 CFR 1022)?					
Wetlands [EO 11990; 10 CFR 1022; 40 CFR 1506.27(b)(3)]?					
Threatened, endangered or candidate species and/or their critical habitat, and other special status (e.g., state-listed) species [16 USC 1531; 40 CFR 1508.21(b)(3)]?					
Prime or unique farmland [7 USC 4201; 7 CFR 658; 40 CFR 1508.27(b)(3)]?					
State or national parks, forests, conservation areas or other areas of recreational, ecological, scenic or aesthetic importance?					
Wild and scenic rivers [16 USC 1271; 40 CFR 1508.27(b)(3)]?					
Natural resources (e.g., timber, range, soils, minerals, fish, wildlife, water, bodies and aquifers)?					

(*Continued*)

TABLE I.3 (CONTINUED)
NEPA Environmental Assessment Checklist

List 1: General	Yes	No	N/A	PAGE	Adequacy Evaluation and Comments
Property of historic, archaeological or architectural significance (including sites on or eligible for the National Register of Historic Places and the National Registry of Natural Landmarks) [16 USC 470; 36 CFR 800; 40 CFR 1508.27(b)(3)]?					
Native American' concerns (16 USC 470; 42 USC 1996)?					
Minority and low-income populations (including a description of their use and consumption of environmental resources) (EO 12898)?					
1.4.2 Does the description of the affected environment provide the necessary information to support the impact analysis, including cumulative impact analysis?					
1.4.3 Does the EA appropriately use incorporation by reference? Is/are the incorporated document(s) up to date?					
1.4.4 If this EA adopts, in whole or in part, a NEPA document prepared by another federal agency, has agency independently evaluated the information?					
1.5.0 ENVIRONMENTAL EFFECTS					
1.5.1 Does the EA identify the potential effects (including cumulative effects) to the following, as identified in question 1.4.1:					
Floodplains (EO 11968; 10 CFR 1022)?					
Wetlands [EO 11990; 10 CFR 1022; 40 CFR 1506.27(b)(3)]?					
Threatened, endangered or candidate species and/ or their critical habitat, and other special status (e.g., state-listed) species [16 USC 1531; 40 CFR 1508.27(b)(3)]?					
Prime or unique farmland [7 USC 4201; 7 CFR 658; 40 CFR 1508.27(b)(3)]?					
State or national parks, forests, conservation areas or other areas of recreational ecological, scenic or aesthetic importance?					
Wild and scenic rivers [16 USC 1271; 40 CFR 1508.27(b)(3)]?					
Natural resources (e.g., timber, range, soils, minerals, fish, wildlife, water bodies, aquifers)?					

(Continued)

TABLE I.3 (CONTINUED)
NEPA Environmental Assessment Checklist

List 1: General	Yes	No	N/A	PAGE	Adequacy Evaluation and Comments
Property of historic, archaeological or architectural significance (including sites on or eligible for the National Registry of Natural Landmarks) [16 USC 470; 36 CFR 800; 40 CFR 1508.27(b)(3)]?					
Native Americans' concerns (16 USC 470; 42 USC 1996)?					
Minority and low-income populations (EO 12898)?					
1.5.2 Does the EA analyze the proposed action:					
For both short-term and long-term effects [40 CFR 1508.27(a)]?					
For both beneficial and adverse impacts [40 CFR 1508.27(b)(1)]?					
For effects on public health and safety [40 CFR 1508.27(b)(2)]?					
For disproportionately high and adverse human health or environmental effects on minority and low-income communities [EO 12898]?					
1.5.3 Do the discussions of environmental impacts include (as appropriate) human health effects, effects of accidents and transportation effects (Recommendations Sec. 6.1)?					
1.5.4 As appropriate, does the EA address the degree to which the possible effects on the human environment may be highly uncertain or involve unique or unknown risks [40 CFR 1508.27(b)(5)]?					
1.5.5 Do the discussions of environmental impacts identify possible indirect and cumulative impacts?					
1.5.6 Does the EA quantify environmental impacts where possible?					
1.5.7 Are all potentially nontrivial impacts identified? Are impacts analyzed using a graded approach – i.e., proportional to their potential significance?					
1.5.8 Does the EA identify all reasonably foreseeable impacts (40 CFR 1508.8)?					
1.5.9 If information related to potential impacts is incomplete or unavailable, does the EA indicate that such information is lacking (40 CFR 1502.22)?					

(Continued)

TABLE I.3 (CONTINUED)
NEPA Environmental Assessment Checklist

List 1: General	Yes	No	N/A	PAGE	Adequacy Evaluation and Comments
1.5.10 Are sufficient data and references presented to allow review of the validity of analysis methods and results?					
1.6.0 OVERALL CONSIDERATIONS/INCORPORATIONS OF NEPA VALUES					
1.6.1 Because conclusions of overall significance will be made in a FONSI or determination to prepare an EIS, are the words 'significant' and 'insignificant' absent from conclusory statements in the EA?					
1.6.2 Do the conclusions regarding potential impacts follow from the information and analyses presented in the EA?					
1.6.3 Does the EA avoid the implication that compliance with regulatory requirements demonstrates the absence of significant environmental effects?					
1.6.4 Are mitigation measures appropriate to the potential impacts identified in the EA [40 CFR 1500.2(f)]?					
1.6.5 Does the EA show that the agency 'has taken a "hard look" at environmental consequences' [*Kleppe v. Sierra Club*, 427 US 390, 410 (1976)]?					
1.7.0 PROCEDURAL CONSIDERATIONS					
1.7.1 Were host states and tribes and, when applicable, the public notified of agency's determination to prepare the EA (10 CFR 1021.301; Policy Statement, Sec. V)? Does the EA address issues known to be of concern to the states, tribes and public?					
1.7.2 Has the EA been made available to the agencies, states, tribes and the public? (10 CFR 1021.301)?					
1.7.3 Have stakeholders including the public been involved to the extent practicable during the preparation of the EA [CEQ (46 FR 18037); 40 CFR 1506.6; 40 CFR 1501.4(b); 10 CFR 1021.301]? Has the Agency proactively sought the involvement of minority and low-income communities in the review and preparation process (EO 12898)?					
1.7.4 Have comments from host states and tribes and, when applicable, the public been addressed (10 CFR 1021.301; Policy Statement, Sec. V)?					

(Continued)

TABLE I.3 (CONTINUED)
NEPA Environmental Assessment Checklist

List 1: General	Yes	No	N/A	PAGE	Adequacy Evaluation and Comments
1.7.5 Is a Floodplain/Wetlands Assessment required and if so, has one been completed? If required, has a Public Notice been published in the Federal Register [10 CFR 1022.14(b)]?					
1.7.6 Does the EA demonstrate adequate consultation with appropriate agencies to ensure compliance with sensitive resource laws and regulations? Are letters of consultation (e.g., SHPO, USFWS) appended (16 USC 1531; 36 CFR 800)?					
1.7.7 Does the EA include a listing of agencies and persons consulted [40 CFR 1508.9(b)]?					
1.8.0 FORMAT, GENERAL DOCUMENT QUALITY, USER-FRIENDLINESS					
1.8.1 Is the EA written precisely and concisely, using plain language and without jargon [10 CFR 1021.301(b)]?					
1.8.2 Is Agency listed as the preparer on the title page of the EA?					
1.8.3 Is the metric system of units used (with English units in parentheses) to the extent possible?					
1.8.4 If scientific notation is used, is an explanation provided?					
1.8.5 Are technical terms defined where necessary [10 CFR 1021.301(b)]?					
1.8.6 Are the units consistent throughout the document?					
1.8.7 If regulatory terms are used, are they consistent with their regulatory definitions?					
1.8.8 Are visual aids used whenever possible to simplify the EA?					
1.8.9 Are abbreviations and acronyms defined the first time they are used?					
1.8.10 Is the use of abbreviations minimized to the extent practical?					
1.8.11 Do the appendices support the content and conclusions contained in the main body of the EA? Is information in the appendix consistent with information in the main body of the EA?					
1.8.12 Is information in tables and figures consistent with information in the text and appendices?					

(Continued)

TABLE I.3 (CONTINUED)
NEPA Environmental Assessment Checklist

List 1: General	Yes	No	N/A	PAGE	Adequacy Evaluation and Comments
1.9.0 KEY TO SUPPLEMENTAL TOPICAL QUESTIONS					
1.9.1 Does the proposed action present potential for impacts on water resources or water quality?					If yes, complete questions in Section 2.1.0
1.9.2 Does the proposed action present potential for impacts related to geology or soils?					If yes, complete questions in Section 2.2.0
1.9.3 Does the proposed action present potential for impacts on air quality?					If yes, complete questions in Section 2.3.0
1.9.4 Does the proposed action present potential for impacts on wildlife or habitat?					If yes, complete questions in Section 2.4.0
1.9.5 Does the proposed action present potential for effects on human health?					If yes, complete questions in Section 2.5.0
1.9.6 Does the proposed action involve transportation?					If yes, complete questions in Section 2.6.0
1.9.7 Does the proposed action involve waste management?					If yes, complete questions in Section 2.7.0
1.9.8 Does the proposed action present potential for impacts on socioeconomic conditions?					If yes, complete questions in Section 2.8.0
1.9.9 Does the proposed action present potential for impacts on historic, archaeological or other cultural sites or properties?					If yes, complete questions in Section 2.9.0

List 2: Supplemental Topics	Yes	No	N/A	Page	Adequacy Evaluation and Comments
2.1.0 WATER RESOURCES AND WATER QUALITY					
2.1.1 Does the EA identify potential effects of the proposed action and alternatives on surface water quantity and quality under both normal operations and accident conditions?					
2.1.2 Does the EA evaluate whether the proposed action or alternatives would be subject to:					
Water quality or effluent standards?					
National Interim Primary Drinking Water Regulations?					
National Secondary Drinking Water Regulations?					
2.1.3 Does the EA state whether the proposed action or alternatives:					
Would include work in, under, over or having an effect on navigable water of the United States?					
Would include the discharge of dredged or fill material into waters of the United States?					

(Continued)

TABLE I.3 (CONTINUED)
NEPA Environmental Assessment Checklist

List 1: General	Yes	No	N/A	PAGE	Adequacy Evaluation and Comments
Would include the deposit of fill material or an excavation that alters or modifies the course, location, condition or capacity of any navigable waters of the United States?					
Would require a Rivers and Harbors Act Section 10 permit or a Clean Water Act (Section 402 or Section 404) permit?					
2.1.3 Does the EA state whether the proposed action or alternatives:					
2.1.4 Does the EA identify potential effects of the proposed action and alternatives on groundwater quantity and quality (including aquifers) under both normal operations and accident conditions?					
2.1.5 Does the EA consider whether the proposed action or alternatives may affect any municipal or private drinking water supplies?					
2.2.0 GEOLOGY AND SOILS					
2.2.1 Does the EA describe and quantify the land area proposed to be altered, excavated or otherwise disturbed? Is this description consistent with other sections (e.g., land use, habitat area)?					
2.2.2 Are issues related to seismicity sufficiently characterized, quantified and analyzed?					
2.2.3 If the action involves disturbance of surface soils, are erosion control measures addressed?					
2.3.0 AIR QUALITY					
2.3.1 Does the EA identify potential effects of the proposed action on ambient air quality under both normal and accident conditions?					
2.3.2 Are potential emissions quantified to the extent practicable (amount and rate of release)?					
2.3.3 Does the EA evaluate potential effects on human health and the environment from exposure to radiation and hazardous chemicals in emissions?					
2.3.4 Does the EA evaluate whether the proposed action and alternatives would:					
Be in compliance with the National Ambient Air Quality Standards?					
Be in compliance with the State Implementation Plan?					
Potentially affect any area designated as Class I under the Clean Air Act?					

(Continued)

TABLE I.3 (CONTINUED)
NEPA Environmental Assessment Checklist

List 1: General	Yes	No	N/A	PAGE	Adequacy Evaluation and Comments
Be subject to New Source Performance Standards?					
Be subject to National Emissions Standards for Hazardous Air Pollutants?					
Be subject to emissions limitations in an Air Quality Control Region?					
2.4.0 WILDLIFE AND HABITAT					
2.4.1 If the EA identifies potential effects of the proposed action and alternatives on threatened or endangered species and/or critical habitat, has the consultation with the USFWS or NMFS been concluded? Does the EA address *candidate* species?					
2.4.2 Are *state*-listed species identified, and if so, are results of state consultation documented?					
2.4.3 Are potential effects (including cumulative effects) analyzed for fish and wildlife other than threatened and endangered species and for habitats other than critical habitats?					
2.4.4 Does the EA analyze the impacts of the proposed action on the biodiversity of the affected ecosystem, including genetic diversity and species diversity?					
2.4.5 Are habitat types identified and estimates provided by type for the amount of habitat lost or adversely affected?					
2.5.0 HUMAN HEALTH EFFECTS					
2.5.1 Have the susceptible populations been identified – i.e., involved workers, noninvolved workers and the public (including minority and low-income communities, as appropriate)?					
2.5.2 Does the EA establish the period of exposure (e.g., 30 years or 70 years) for exposed workers and the public?					
2.5.3 Does the EA identify all potential routes of exposure?					
2.5.4 When providing quantitative estimates of impacts, does the EA use current dose-to-risk conversion factors that have been adopted by cognizant health and environmental agencies?					

(Continued)

TABLE I.3 (CONTINUED)
NEPA Environmental Assessment Checklist

List 1: General	Yes	No	N/A	PAGE	Adequacy Evaluation and Comments
2.5.5 When providing quantitative estimates of health effects due to radiation exposure, are collective effects expressed in estimated numbers of fatal cancers, and are maximum individual effects expressed as the estimated maximum probability of death of an individual?					
2.5.6 Does the EA describe assumptions used in the health effects analysis and the basis for health effects calculations?					
2.5.7 As appropriate, does the EA analyze radiological impacts under normal operating conditions for:					
Involved workers:					
Collective dose?					
Maximum individual?					
Latent cancer fatalities?					
Uninvolved workers:					
Collective dose?					
Maximum individual?					
Latent cancer fatalities?					
Public:					
Collective dose?					
Maximum individual?					
Latent cancer fatalities?					
2.5.8 Does the EA identify a spectrum of potential accident scenarios that could occur over the life of the proposed action?					
2.5.9 As appropriate, does the EA analyze radiological impacts under accident conditions for:					
Involved workers:					
Collective dose?					
Maximum individual?					
Latent cancer fatalities?					
Uninvolved workers:					
Collective dose?					
Maximum individual?					
Latent cancer fatalities?					
Public:					
Collective dose?					
Maximum individual?					
Latent cancer fatalities?					

(Continued)

TABLE I.3 (CONTINUED)
NEPA Environmental Assessment Checklist

List 1: General	Yes	No	N/A	PAGE	Adequacy Evaluation and Comments
2.5.10 Are nonradiological impacts (e.g., chemical exposures) addressed for both routine and accident conditions?					
2.6.0 TRANSPORTATOIN					
2.6.1 If transport of hazardous or radioactive waste or materials is part of the proposed action, or if transport is a major factor, are the potential effects analyzed (including to a site, on-site and from a site)?					
2.6.2 Does the EA analyze all reasonably foreseeable transportation links (e.g., overland transport, port transfer, marine, transport, global commons) (EO 12114)?					
2.6.3. Does the EA avoid relying exclusively on statements that transportation will be in accordance with all applicable state and federal regulations and requirements?					
2.6.4 Does the EA address both routine transportation and reasonably foreseeable accidents?					
2.6.5 Are the estimation methods used for assessing radiological impacts of transportation defensible?					
2.6.6 Does the EA address the annual, total and cumulative impacts of all Agency and non-Agency transportation on specific routes associated with the proposed action?					
2.7.0 WASTE MANAGEMENT AND WASTE MINIMIZATION					
2.7.1 Are pollution prevention and waste minimization practices applied in the proposed action and alternatives (e.g., is pollution prevented or reduced at the source when feasible; would waste products be recycled when feasible; are by-products that cannot be prevented or recycled treated in an environmentally safe manner when feasible; is disposal only used as a last resort)?					
2.7.2 If waste would be generated, does the EA examine the human health effects and environmental impacts of managing that waste, including waste generated during decontaminating and decommissioning?					

(Continued)

TABLE I.3 (CONTINUED)
NEPA Environmental Assessment Checklist

List 1: General	Yes	No	N/A	PAGE	Adequacy Evaluation and Comments
2.7.3 Are waste materials characterized by type and estimated quantity, where possible?					
2.7.4 Does the EA identify RCRA/CERCLA issues related to the proposed action and alternatives?					
2.7.5 Does the EA establish whether the proposed action and alternatives would be in compliance with federal or state laws and guidelines affecting the generation, transportation, treatment, storage or disposal of hazardous and other waste?					
2.8.0 SOCIOECONOMIC CONSIDERATIONS					
2.8.1 Does the EA consider potential effects on land use patterns, consistency with applicable land use plans and compatibility of nearby uses?					
2.8.2 Does the EA consider potential economic impacts, such as effects on jobs and housing, particularly in regard to disproportionate adverse effects on minority and low-income communities?					
2.8.3 Does the EA consider potential effects on public water and wastewater services, storm water management, community services and utilities?					
2.8.4 Does the EA evaluate potential noise effects of the proposed action and the application of community noise level standards?					
2.9.0 CULTURAL RESOURCES					
2.9.1 Was the SHPO consulted?					
2.9.2 Was an archaeological survey conducted?					
2.9.3 Does the EA include a provision for mitigation in the event when unanticipated archaeological materials are encountered?					

TABLE I.4
NEPA Environmental Impact Checklist

List 1: General	Yes	No	N/A	EIS Page	Adequacy Evaluation and Comments
1.1.0 COVER SHEET					
1.1.1 Does the cover sheet include:					
A list of responsible agencies, including the lead agency and any cooperating agencies?					
The title of the proposed action and its location (state(s), county(ies), other jurisdictions)?					
The name(s), address(es) and telephone number(s) of a person (or persons) to contact for further information (on the general Agency NEPA process and on the specific EIS)?					
The EIS designation as draft, final or supplemental?					
A one-paragraph abstract of the EIS?					
For a draft EIS, the date by which comments must be received? (40 CFR 1502.11)					
1.1.2 Is the cover sheet one page in length? (40 CFR 1502.11)					
1.2.0 SUMMARY					
1.2.1 Does the summary describe:					
The underlying purpose and need for agency action? the proposed action?					
Each of the alternatives?					
The preferred alternative, if any?					
The principal environmental issues analyzed and results?					
1.2.2 Does the summary highlight key differences among the alternatives?					
1.2.3 Does the summary stress:					
The major conclusions?					
Areas of controversy (including issues raised by agencies and the public)?					
The issues to be resolved (including the choice among alternatives)?					
1.2.4 Are the discussions in the Summary consistent with the EIS text or appendices?					
1.2.5 Does the summary adequately and accurately summarize the EIS? (40 CFR 1502.12)					
1.3.0 PURPOSE AND NEED FOR ACTION					
1.3.1 Does the EIS specify the underlying purpose and need to which DOE is responding in proposing the alternatives including the proposed action?					

(Continued)

TABLE I.4 (CONTINUED)
NEPA Environmental Impact Checklist

List 1: General	Yes	No	N/A	EIS Page	Adequacy Evaluation and Comments
(40 CFR 1502.13)					
1.3.2 Does the statement of purpose and need relate to the broad requirement or desire for *Agency* action, and not to the need for one specific proposal or the need for the EIS?					
1.3.3 Does the statement of purpose and need adequately explain the problem or opportunity to which this Agency is responding?					
1.3.4 Is the statement of purpose and need written so that it (a) does not inappropriately narrow the range of reasonable alternatives, or (b) is not too broadly defined as to make the number of alternatives virtually limitless?					
1.4.0 DESCRIPTION OF THE PROPOSED ACTION AND ALTERNATIVES					
1.4.1 Does the EIS clearly describe the proposed action and alternatives?					
1.4.2 Is the proposed action described in terms of the Agency action to be taken (even a private action that has been federalized or enabled by funding)?					
1.4.3 Does the proposed action exclude elements that are more appropriate to the statement of purpose and need?					
1.4.4 Does the EIS identify the range of reasonable alternatives that satisfy the agency's purpose and need?					
1.4.5 Does the EIS 'rigorously explore and objectively evaluate' all reasonable alternatives that encompass the range to be considered by the decision-maker? [40 CFR 1502.14(a)]					
1.4.6a For a *draft* EIS, does the document indicate whether a preferred alternative(s) exists, and, if so, is it identified? [40 CFR 1502.14(e)]					
1.4.6b For a *final* EIS, is the preferred alternative identified? [40 CFR 1502.14(e)]					
1.4.7 Does the EIS include the no action alternative? [40 CFR 1502.14(d)]					
1.4.8 Is the no action alternative described in sufficient detail so that its scope is clear and potential impacts can be identified?					
1.4.9 Does the no action alternative include a discussion of the legal ramifications of no action, if appropriate?					
1.4.10 As appropriate, does the EIS identify and analyze reasonable technology, transportation and siting alternatives, including those that could occur off-site?					

(Continued)

TABLE I.4 (CONTINUED)
NEPA Environmental Impact Checklist

List 1: General	Yes	No	N/A	EIS Page	Adequacy Evaluation and Comments
1.4.11 Does the EIS include reasonable alternatives outside of the Agency's jurisdiction? [40 CFR 1502.14(c)]					
1.4.12 For alternatives that were eliminated from detailed study, including those that appear obvious or were identified by the public, does the EIS explain fully and objectively why they were found to be unreasonable? [40 CFR 1502.14(a)]					
1.4.13 For each alternative analyzed in detail (including the no action alternative), is the depth of analysis approximately the same, allowing reviewers to evaluate their comparative merits? [40 CFR 1502.14(b)]					
1.4.14 Are the proposed action and alternatives described in sufficient detail so that potential impacts can be identified?					
1.4.15 Are all phases of the proposed action and alternatives described (e.g., construction, operation and postoperation/ decommissioning)?					
1.4.16 Are environmental releases associated with the proposed action and alternatives quantified, including both the rates and durations?					
1.4.17 As appropriate, are mitigation measures included in the description of the proposed action and alternatives? [40 CFR1502.14(f)]					
1.4.18 Are cost-effective waste minimization and pollution prevention activities included in the description of the proposed action and alternatives?					
1.4.19 As appropriate, are environmentally and economically beneficial landscape practices included in the description of the proposed action and alternativess (60 FR 40837)					
1.4.20 Are the descriptions of the proposed action and alternatives written broadly enough to encompass future modifications?					
1.4.21 Does the proposed action comply with CEQ regulations for interim actions? (40 CFR 1506.1)					
1.4.22 Does the EIS take into account relationships between the proposed action and other actions to be taken by the agency in order to avoid improper segmentation?					
1.5.0 DESCRIPTION OF THE AFFECTED ENVIRONMENT					
1.5.1 Does the EIS succinctly describe the environment of the area(s) to be affected or created by the proposed action and alternatives? (40 CFR 1502.15)					

(Continued)

TABLE I.4 (CONTINUED)
NEPA Environmental Impact Checklist

List 1: General	Yes	No	N/A	EIS Page	Adequacy Evaluation and Comments
1.5.2 Does the EIS *identify either the presence or absence* of the following within the area potentially affected by the proposed action and alternatives:					
Floodplains? (EO 11988; 10 CFR 1022)					
Wetlands? [EO 11990; 10 CFR 1022; 40 CFR 1508.27(b)(3)]					
Threatened, endangered or candidate species and/or their critical habitat, and other special status (e.g., state-listed) species? [16 USC 1531; 40 CFR 1508.27(b)(9)]					
Prime or unique farmland? [7 USC 4201; 7 CFR 658; 40 CFR 1508.27(b)(3)]					
State or national parks, forests, conservation areas or other areas of recreational, ecological, scenic or aesthetic importance? [40 CFR 1508.27(b)(3)]					
Wild and scenic rivers? [16 USC 1271; 40 CFR 1508.27(b)(3)]					
Natural resources (e.g., timber, range, soils, minerals, fish, migratory birds, wildlife, water bodies, aquifers)? (40 CFR 1508.8)					
property of historic, archaeological or architectural significance (including sites on or eligible for the National Register of Historic Places and the National Registry of Natural Landmarks)? E[O 11593; 16 USC 470; 36 CFR 800; 40 CFR 1508.27(b)(3) and (8)]					
Native Americans' concerns? [EO 13007; 25 USC 3001; 16 USC 470; 42 USC 1996]					
minority and low-income populations (including a description of their use and consumption of environmental resources)? [EO 12898]					
1.5.3 Does the description of the affected environment provide the necessary information to support the impact analysis, including cumulative impact analysis? (40 CFR 1502.15; Recommendations, p.14)					
1.5.4 Are the descriptions of the affected environment substantially consistent with current site baseline studies (e.g., descriptions of plant communities, wildlife habitat and cultural resources)?					
1.5.5 Is the discussion appropriately limited to information that is directly related to the scope of the proposed action and alternatives? (40 CFR 1502.15)					

(Continued)

TABLE I.4 (CONTINUED)
NEPA Environmental Impact Checklist

List 1: General	Yes	No	N/A	EIS Page	Adequacy Evaluation and Comments
1.5.6 Is the extent of each component of the affected environment appropriately described with respect to potential impacts (e.g., the affected environment for transportation impacts may be more extensive than that for groundwater impacts)?					
1.5.7 Does the EIS avoid useless bulk and verbose descriptions of the affected environment and concentrate on important issues? (40 CFR 1502.15)					
1.6.0 ENVIRONMENTAL EFFECTS					
1.6.1 Does the EIS adequately identify the direct and the *indirect* impacts of the proposed action and alternatives and discuss their significance? [40 CFR 1502.16(a) and (b)]					
1.6.2 Does the EIS adequately analyze both short-tterm and long-term effects?					
1.6.3 Does the EIS analyze both beneficial and adverse impacts? [40 CFR 1508.27(b)(1)]					
1.6.4 Does the EIS discuss reasonably foreseeable impacts of cumulative actions with regard to both the proposed action and alternatives? [40 CFR 1508.25(a)(2)]					
1.6.5 Does the EIS discuss the potential direct, indirect and cumulative effects to the following, as identified in Question 1.5.2:					
Floodplains? (EO 11988; 10 CFR 1022)					
Wetlands? [EO 11990; 10 CFR 1022; 40 CFR 1508.27(b)(3)]					
Threatened, endangered or candidate species and/or their critical habitat, and other special status (e.g., state-listed) species? [16 USC 1531; 40 CFR 1508.27(b)(9)]					
Prime or unique farmland? [7 USC 4201; 7 CFR 658; 40 CFR 1508.27(b)(3)]					
State or national parks, forests, conservation areas or other areas of recreational, ecological, scenic or aesthetic importance? [40 CFR 1508.27(b)(3)]					
Wild and scenic rivers? [16 USC 1271; 40 CFR 1508.27(b)(3)]					
Natural resources (e.g., timber, range, soils, minerals, fish, migratory birds, wildlife, water bodies, aquifers)? (40 CFR 1508.8)					

(Continued)

TABLE I.4 (CONTINUED)
NEPA Environmental Impact Checklist

List 1: General	Yes	No	N/A	EIS Page	Adequacy Evaluation and Comments
Property of historic, archaeological or architectural significance (including sites on or eligible for the National Register of Historic Places and the National Registry of Natural Landmarks)? E[O 11593; 16 USC 470; 36 CFR 800; 40 CFR 1508.27(b)(3) and (8)]					
Native Americans' concerns? (EO 13007; 25 USC 3001; 16 USC 470; 42 USC 1996)					
Minority and low-income populations to the extent that such effects are disproportionately high and adverse? (EO 12898)					
1.6.6 Does the EIS discuss:					
Possible conflicts with land use plans, policies or controls? [40 CFR 1502.16(c)]					
Energy requirements and conservation potential of various alternatives and mitigation measures? [40 CFR 1502.16(e)]					
Natural or depletable resource requirements and conservation potential of the proposed action and alternatives? [40 CFR 1502.16(f)]					
1.6.6 Does the EIS discuss:					
Urban quality, historic and cultural resources and the design of the built environment, including the reuse and conservation potential of the proposed action and alternatives? [40CFR 1502.16(g)]					
The means to mitigate adverse impacts? [40 CFR 1502.16(h)]					
1.6.7 Does the EIS discuss:					
Any unavoidable, adverse environmental effects?					
The relationship between short-term uses of the environment and long-term productivity?					
Any irreversible or irretrievable commitments of resources? (40 CFR 1502.16)					
1.6.8 Do the discussions of environmental impacts include (as appropriate):					
Human health effects? (Recommendations, p2. 0)					
Effects of accidents? (Recommendations, p2. 7)					
Transportation effects?					
1.6.9 Does the EIS discuss the potential effects of released pollutants, rather than just identifying the releases?					
1.6.10 Does the EIS avoid presenting a description of severe impacts (e.g., from accidents), without also describing the likelihood/probability of such impacts occurring?					

(Continued)

TABLE I.4 (CONTINUED)
NEPA Environmental Impact Checklist

List 1: General	Yes	No	N/A	EIS Page	Adequacy Evaluation and Comments
1.6.11 Are the methodologies used for impact assessment generally accepted/recognized in the scientific community? (40 CFR 1502.22 and 1504.24)					
1.6.12 Does the EIS quantify environmental impacts where possible?					
1.6.13 Are impacts analyzed using a graded approach; i.e., proportional to their potential significance?					
1.6.14 Does the EIS avoid presenting bounding impact estimates that obscure differences among alternatives?					
1.6.15 Are sufficient data and references presented to allow validation of analysis methods and results?					
1.6.16a If information related to significant adverse effects is incomplete or unavailable, does the EIS state that such information is lacking?					
1.6.16b If this information is essential to a choice among alternatives and the costs of obtaining it are not exorbitant, is the information included?					
1.6.16c If this information cannot be obtained, does the EIS include: (1) a statement that the information is incomplete or unavailable, (2) the relevance of the information to evaluating significant effects, (3) a summary of credible scientific evidence and (4) an evaluation based on theoretical approaches? (40 CFR 1502.22)					
1.6.17 As appropriate, does the EIS identify important sources of uncertainty in the analyses and conclusions?					
1.7.0 OVERALL CONSIDERATIONS/INCORPORATIONOF NEPA VALUES					
1.7.1 Does the EIS identify *all reasonably foreseeable* impacts? (40 CFR 1508.8)					
1.7.2 Do the conclusions regarding potential impacts follow from the information and analyses presented in the EIS?					
1.7.3 Does the EIS avoid the implication that compliance with regulatory requirements demonstrates the absence of environmental effects?					
1.7.4 To the extent possible, does the EIS assess reasonable alternatives and identify measures to restore and enhance the environment and avoid or minimize potential adverse effects? [40 CFR 1500.2(f)]					
1.7.5 Does the EIS identify best management practices associated with the proposed action or with mitigation measures that would help avoid or minimize environmental disturbance, emissions and other adverse effects?					

(*Continued*)

TABLE I.4 (CONTINUED)
NEPA Environmental Impact Checklist

List 1: General	Yes	No	N/A	EIS Page	Adequacy Evaluation and Comments
1.7.6 Does the EIS avoid the appearance of justifying decisions that have already been made? (40 CFR 1502.5)					
1.7.7 Are all assumptions conservative, and are analyses and methodologies generally accepted/recognized by the scientific community? (40 CFR 1502.22 and 1502.24)					
1.7.8 Does the EIS show that the agency 'has taken a "hard look" at environmental consequences?' [*Kleppe v. Sierra Club*, 427 US 390, 410 (1976)]					
1.7.9 Does the EIS present the potential environmental effects of the proposal and the alternatives in comparative form, sharply defining the issues and providing a clear basis for choice? (40 CFR 1502.14)					
1.8.0 FORMAT, GENERAL DOCUMENT QUALITY, USER-FRIENDLINESS					
1.8.1 Is the EIS written precisely and concisely, using plain language and defining any technical terms that must be used? [10 CFR 1021.301(b)]					
1.8.2 Is information in tables and figures consistent with information in the text and appendices?					
1.8.3 Is the metric system of units used (with English units in parentheses) to the extent possible?					
1.8.4 Are the units consistent throughout the document?					
1.8.5 Are technical terms defined, using plain language, where necessary? [10 CFR 1021.301(b)]					
1.8.6 If scientific notation is used, is an explanation provided? (Recommendations, p.35)					
1.8.7 If regulatory terms are used, are they consistent with their regulatory definitions?					
1.8.8 Does the EIS use conditional language (i.e., 'would' rather than 'will') in describing the proposed action and alternatives and their potential consequences?					
1.8.9 Are graphics and other visual aids used whenever possible to simplify the EIS?					
1.8.10 Are abbreviations and acronyms defined the first time they are used?					
1.8.11 Is the use of abbreviations and acronyms minimized to the extent practical?					
1.8.12 Does the EIS make appropriate use of appendices (e.g., for material prepared in connection with the EIS and related environmental reviews, substantiating material official communications and descriptions of methodologies)? (40 CFR 1502.18 and 1502.24)					

(*Continued*)

TABLE I.4 (CONTINUED)
NEPA Environmental Impact Checklist

List 1: General	Yes	No	N/A	EIS Page	Adequacy Evaluation and Comments
1.8.13 Do the appendices support the content and conclusions contained in the main body of the EIS?					
1.8.14 Is there a discussion of the relationship between this EIS and related DOE NEPA documents?					
1.8.15 Is the issue date (month and year of approval) on the cover?					
1.9.0 OTHER REGULATORY REQUIREMENTS					
1.9.1 Unless there is a compelling reason to do otherwise, does the EIS include:					
Table of contents? Index?					
List of agencies, organizations and persons to whom copies of the EIS were sent? (40 CFR 1502.10)					
1.9.2 Does the EIS identify all federal permits, licenses and other entitlements that must be obtained in implementing the proposal? [40 CFR 1502.25(b)]					
1.9.3. Does the EIS identify methodologies used in the analyses, include references to sources relied upon for conclusions and provide documentation or references to documentation for methodologies? (40 CFR 1502.24)					
1.9.4 If a cost-benefit analysis has been prepared, is it incorporated by reference or appended to the EIS? (40 CFR 1502.23)					
1.9.5 If this EIS adopts, in whole or in part, a NEPA document prepared by another federal agency, has Agency independently evaluated the information? (40 CFR 1506.3)					
1.9.6 Does the EIS appropriately use incorporation by reference, i.e.:					
Is the information up to date?					
Is the information summarized in the EIS?					
Are cited references publicly available? (40 CFR 1502.21); Recommendations, pp.14 and 37)					
1.9.7 Does the EIS contain a list of preparers and their qualifications? (40 CFR 1502.17)					
1.9.8 Does the EIS include a contractor disclosure statement? [40 CFR 1506.5(c); 10 CFR 1021.310]					
1.9.9 Is Agency listed as the preparer on the title page of the EIS and has Agency evaluated all information and accepted responsibility for the contents? (40 CFR 1506.5)					

(Continued)

TABLE I.4 (CONTINUED)
NEPA Environmental Impact Checklist

List 1: General	Yes	No	N/A	EIS Page	Adequacy Evaluation and Comments
1.10.0 PROCEDURAL CONSIDERATIONS					
1.10.1 Did Agency notify the host state and host tribe, and other affected states and tribes, of the determination to prepare the EIS?					
[10 CFR 1021.301(c)]					
1.10.2 Did Agency publish a Notice of Intent in the Federal Register, allowing reasonable time for public comment? [10 CFR 1021.311(a) and 40 CFR 1501.7]					
1.10.3 Is a Floodplain/Wetlands Assessment required, and if so, has a Notice of Involvement been published in the Federal Register? (10 CFR 1022.14)					
1.10.4 In addition to EPA's Notice of Availability, did Agency otherwise publicize the availability of the draft EIS, focusing on potentially interested or affected persons? (40 CFR 1506.6)					
1.10.5 Did Agency actively seek the participation of low-income and minority communities in the preparation and review of the EIS? (EO 12898; Effective Public Participation guidance, p.11)					
1.10.6 Is the administrative record for this EIS being maintained contemporaneously, and does it provide evidence that Agency considered all relevant issues?					
1.10.7 To the fullest extent possible, have other environmental review and consultation requirements been integrated with NEPA requirements? (40 CFR 1502.25)					
1.11.0 DRAFT EIS CONSIDERATIONS					
1.11.1 Has Agency considered scoping comments from other agencies and the public? [10 CFR 1021.311(e)]					
1.11.2 Does the draft EIS demonstrate that Agency considered possible connected actions, cumulative actions and similar actions? [40 CFR 1508.25(a)]					
1.11.3 If the draft EIS identifies a preferred alternative(s), does the document present the criteria and selection process? [40 CFR 1502.14(e)]					
1.11.4a Does the draft EIS demonstrate adequate consultation with appropriate agencies to ensure compliance with sensitive resource laws and regulations?					
1.11.4b Does the document contain a list of agencies and persons consulted?					

(Continued)

TABLE I.4 (CONTINUED)
NEPA Environmental Impact Checklist

List 1: General	Yes	No	N/A	EIS Page	Adequacy Evaluation and Comments
1.11.4c Are letters of consultation (e.g., SHPO, USFWS) appended? (40 CFR 1502.25)					
1.12.0 FINAL EIS CONSIDERATIONS					
1.12.1 Does the final EIS discuss at appropriate points responsible opposing views not adequately addressed in the draft EIS and indicate Agency's responses to the issues raised? [40 CFR 1502.9 (b)]					
1.12.2a Is the preferred alternative identified? [40 CFR 1502.14(e)]					
1.12.2b Does the document present the criteria and selection process for the preferred alternative?					
1.12.3 Does the final EIS demonstrate, through appropriate responses that all substantive comments from other agencies, organizations and the public were objectively considered, both individually and cumulatively (i.e., by modifying the alternatives, developing new alternatives, modifying and improving the analyses, making factual corrections or explaining why the comments do not warrant agency response)? (40 CFR 1503.4)					
1.12.4 Are all substantive comments (or summaries thereof) and Agency responses included in the final EIS? [40 CFR 1503.4(b)]					
1.12.5 Are any changes to the draft EIS clearly marked or otherwise identified in the final EIS?					
1.12.6 Is the final EIS suitable for filing with EPA, i.e.:					
Does it have a new cover sheet?					
Does it include comments and responses?					
Does it include any revisions or supplements to the draft? (40 CFR 1503.4 and 1506.9)					
2.1.0 WATER RESOURCES AND WATER QUALITY					
2.1.1 Does the EIS discuss potential effects of the proposed action and alternatives:					
On surface water quantity: Under normal operations? Under accident conditions?					
On surface water quality: Under normal operations? Under accident conditions?					
2.1.2 Does the EIS assess the effect of the proposed action and alternatives on the quantity, quality, location and timing of storm water runoff (e.g., will new impervious surfaces create a need for storm water management or pollution controls)?					

(Continued)

TABLE I.4 (CONTINUED)
NEPA Environmental Impact Checklist

List 1: General	Yes	No	N/A	EIS Page	Adequacy Evaluation and Comments
2.1.3 Would the proposed action or alternatives require a storm water discharge permit?					
2.1.4 Does the EIS evaluate whether the proposed action or alternatives would be subject to:					
Water quality or effluent standards?					
National Primary Drinking Water Regulations?					
National Secondary Drinking Water Regulations?					
2.1.5 Does the EIS state whether the proposed action or alternatives:					
Would include work in, under, over or having an effect on navigable waters of the United States?					
Would include the discharge of dredged or fill material into waters of the United States?					
Would include the deposit of fill material or an excavation that alters or modifies the course, location, condition or capacity of any navigable waters of the United States?					
Would require a Rivers and Harbors Act (Section 10) permit or a Clean Water Act (Section 402 or Section 404) permit?					
Would require a determination under the Coastal Zone Management Act? If so, is such determination included in the draft EIS?					
2.1.6 Does the EIS discuss potential effects of the proposed action and alternatives:					
On groundwater quantity:					
Under normal operations?					
Under accident conditions?					
On groundwater quality:					
Under normal operations?					
Under accident conditions?					
2.1.7 Does the EIS consider whether the proposed action or alternatives may affect any municipal or private drinking water supplies?					
2.1.8 Does the EIS evaluate the incremental effect of effluents associated with the proposed action and alternatives in terms of cumulative water quality conditions?					
2.1.9 If the proposed action may involve a floodplain, does the document discuss alternative actions to avoid or minimize impacts and preserve floodplain values?					

(Continued)

TABLE I.4 (CONTINUED)
NEPA Environmental Impact Checklist

List 1: General	Yes	No	N/A	EIS Page	Adequacy Evaluation and Comments
2.2.0 GEOLOGY AND SOILS					
2.2.1 Does the EIS describe and quantify the land area proposed to be altered, excavated or otherwise disturbed?					
2.2.2 Is the description of the disturbed area consistent with other sections (e.g., land use, habitat area)?					
2.2.3 Are issues related to seismicity sufficiently characterized, quantified and analyzed?					
2.2.4 If the action involves disturbance of surface soils, are appropriate best management practices (e.g., erosion control measures) discussed?					
2.2.5 Have soil stability and suitability been adequately discussed?					
2.2.6 Does the EIS consider whether the proposed action may disturb or cause releases of preexisting contaminants or hazardous substances in the soil?					
2.3.0 AIR QUALITY					
2.3.1 Does the EIS discuss potential effects of the proposed action on ambient air quality: Under normal operations? Under accident conditions?					
2.3.2 Are potential emissions quantified to the extent practicable (amount and rate of release)?					
2.3.3 Does the EIS evaluate potential effects to human health and the environment from exposure to radiation emissions?					
2.3.4 Does the EIS evaluate potential effects to human health and the environment from exposure to hazardous chemical emissions?					
2.3.5 When applicable, does the EIS evaluate whether the proposed action and alternatives would: Be in compliance with the National Ambient Air Quality Standards?					
Conform to the State Implementation Plan?					
Potentially affect any area designated as Class I under the Clean Air Act?					
Be subject to New Source Performance Standards?					
Be subject to National Emissions Standards for Hazardous Air Pollutants?					
Be subject to emissions limitations in an Air Quality Control Region?					
2.3.6 Does the EIS evaluate the incremental effect of emissions associated with the proposed action and alternatives in terms of cumulative air quality?					

(Continued)

TABLE I.4 (CONTINUED)
NEPA Environmental Impact Checklist

List 1: General	Yes	No	N/A	EIS Page	Adequacy Evaluation and Comments
2.4.0 WILDLIFE AND HABITAT					
2.4.1 If the EIS identifies potential effects of the proposed action and alternatives on threatened or endangered species and/or critical habitat, has consultation with the USFWS or NMFS been concluded?					
2.4.2 Does the EIS discuss *candidate* species?					
2.4.3 Are *state*-listed species identified, and if so, are results of state consultation documented?					
2.4.4 Are potential effects (including cumulative effects) analyzed for fish and wildlife other than threatened and endangered species and for habitats other than critical habitat?					
2.4.5 Does the EIS analyze the impacts of the proposed action on the biodiversity of the affected ecosystem, including genetic diversity and species diversity?					
2.4.6 Are habitat types identified and estimates provided by type for the amount of habitat lost or adversely affected?					
2.4.7 Does the EIS consider measures to protect, restore and enhance wildlife and habitat?					
2.5.0 HUMAN HEALTH EFFECTS					
2.5.1 Have the following potentially affected populations been identified:					
Involved workers?					
Noninvolved workers?					
The public?					
Minority and low-income communities (as appropriate)? (EO 12898)					
2.5.2 Does the EIS establish the period of exposure (e.g., 30 years or 70 years) for exposed workers and the public?					
2.5.3 Does the EIS identify all potential routes of exposure?					
2.5.4 When providing quantitative estimates of impacts, does the EIS use current dose-to-risk conversion factors that have been adopted by cognizant health and environmental agencies?					
2.5.5 When providing quantitative estimates of health effects due to radiation exposure, are collective effects expressed in estimated numbers of fatal cancers or cancer incidences?					
2.5.6 Are maximum individual effects expressed as the estimated maximum probability of fatality or cancer incidences for an individual?					

(*Continued*)

TABLE I.4 (CONTINUED)
NEPA Environmental Impact Checklist

List 1: General	Yes	No	N/A	EIS Page	Adequacy Evaluation and Comments
2.5.7 Does the EIS describe assumptions used in the health effects analysis and the basis for health effects calculations?					
2.5.8 As appropriate, does the EIS analyze radiological impacts under *normal operating conditions* for:					
Involved workers:					
Population dose and corresponding latent cancer fatalities					
Maximum individual dose and corresponding cancer risk					
Noninvolved workers:					
Population dose and corresponding latent cancer fatalities					
Maximum individual dose and corresponding cancer risk					
Public:					
Population dose and corresponding latent cancer fatalities					
Maximum individual dose and corresponding cancer risk					
2.5.9 Does the EIS identify a reasonable spectrum of potential accident scenarios that could occur over the life of the proposed action, including the maximum reasonably foreseeable accident?					
2.5.10 Does the EIS identify failure scenarios from both natural events (e.g., tornadoes, earthquakes) and human error (e.g., forklift accident)?					
2.5.11 As appropriate, does the EIS analyze radiological impacts under *accident conditions* for:					
Involved workers:					
Population dose and corresponding latent cancer fatalities					
Maximum individual dose and corresponding cancer risk					
Noninvolved workers:					
Population dose and corresponding latent cancer fatalities					
Maximum individual dose and corresponding cancer risk					
2.5.13 Does the EIS discuss toxic and carcinogenic health effects from exposure to hazardous chemicals:					
For involved workers?					
For noninvolved workers?					
For the public?					
Under routine operations?					
Under accident conditions?					
2.5.14 Does the EIS adequately consider physical safety issues for involved and noninvolved workers?					

(Continued)

TABLE I.4 (CONTINUED)
NEPA Environmental Impact Checklist

List 1: General	Yes	No	N/A	EIS Page	Adequacy Evaluation and Comments
2.6.0 TRANSPORTATION					
2.6.1 If transportation of hazardous or radioactive waste or materials is part of the proposed action, or if transportation is a major factor, are the potential effects analyzed (including to a site, on-site and from a site)?					
2.6.2 Does the EIS analyze all reasonably foreseeable transportation links (e.g., overland transport, port transfer, marine transport, global commons)? (EO12114)					
2.6.3 Does the EIS avoid relying exclusively on statements that transportation will be in accordance with all applicable state and federal regulations and requirements?					
2.6.4 Does the EIS discuss routine and reasonably foreseeable transportation accidents?					
2.6.5 Are the estimation methods used for assessing radiological impacts of transportation among those generally accepted/recognized within the scientific community?					
2.6.6 Does the EIS discuss the annual, total and cumulative impacts of all Agency and non-Agency transportation, to the extent such transportation can be estimated, on specific routes associated with the proposed action?					
2.6.7 Have transportation analyses adequately considered potential disproportionately high and adverse impacts on minority and low-income populations? (EO 12898)					
2.7.0 WASTE MANAGEMENT AND WASTE MINIMIZATION					
2.7.1 Are pollution prevention and waste minimization practices applied in the proposed action and alternatives (e.g., is pollution prevented or reduced at the source when feasible; would waste products be recycled when feasible; are by-products that cannot be prevented or recycled treated in an environmentally safe manner when feasible; is disposal only used as a last resort)?					
2.7.2 If waste would be generated, does the EIS examine the human health effects and environmental impacts of managing that waste, including waste generated during decontaminating and decommissioning?					
2.7.3 Are waste materials characterized by type and estimated quantity, where possible?					
2.7.4 Does the EIS identify RCRA/CERCLA issues related to the proposed action and alternatives?					

(Continued)

TABLE I.4 (CONTINUED)
NEPA Environmental Impact Checklist

List 1: General	Yes	No	N/A	EIS Page	Adequacy Evaluation and Comments
2.7.5 Does the EIS establish whether the proposed action and alternatives would be in compliance with federal or state laws and guidelines affecting the generation, transportation, treatment, storage or disposal of hazardous and other waste?					
2.8.0 SOCIOECONOMIC CONSIDERATIONS					
2.8.1 Does the EIS consider potential direct, indirect and cumulative effects on:					
Land use patterns?					
Consistency with applicable land use plans, including site comprehensive plans; and any special designation lands (e.g., farmlands, parks, wildlife conservation areas)?					
Compatibility of nearby uses?					
2.8.2 Does the EIS consider possible changes in the local population due to the proposed action?					
2.8.3 Does the EIS consider potential economic impacts, such as effects on jobs and housing?					
2.8.4 Does the EIS consider potential effects on public water and wastewater services, storm water management and community services and utilities?					
2.8.5 Does the EIS evaluate potential noise effects of the proposed action and the application of community noise level standards?					
2.8.6 Does the EIS state whether the proposed action and alternatives could have disproportionately high and adverse impacts on minority or low-income populations? (EO 12898)					
2.9.0 CULTURAL RESOURCES					
2.9.1 Was the State Historic Preservation Officer consulted?					
2.9.2 Was a cultural resources survey conducted for both archaeological and historical resources, including historic Cold War properties (while maintaining confidentiality by not disclosing locations for sensitive sites)?					
2.9.3 Does the EIS discuss potential access conflicts and other adverse impacts on Native American sacred sites (while maintaining confidentiality by not disclosing locations)? (EO 13007)					
2.9.4 Does the EIS include a provision for mitigation in the event unanticipated archaeological materials (e.g., sites or artifacts) are encountered?					
2.9.5 Does the EIS address consistency of the proposed action with any applicable or proposed cultural resources management plan?					

APPENDIX J
Toolkit for CERCLA

This appendix provides a checklist intended to assist compliance managers and personnel performing a prescreening evaluation under CERCLA. This tool is provided for instructional purpose and guidance only and is not intended to represent an all-inclusive collection of CERCLA regulatory requirements. The tool provided in this appendix are adapted from an EPA checklist and guidance. The example checklist provided in this appendix is a

Pre-CERCLA Screening Checklist/Decision Form.

Pre-CERCLA Screening Checklist/Decision Form

This form is used in conjunction with a site map and any additional information required by the EPA Region to document completion of a Pre-CERCLA Screening (PCS). The form includes a decision on whether a site should be added to the Superfund program's active site inventory for further investigation.

Region: _____ State/Territory: _____ Tribe: _____

EPA ID No. (If Available)

Site Name: _____

Other Site
Name(s): _____

Site _____
                                              (Street)

Congressional   (City)   (State/Terr.)   (County)   (Zip+4)   (No Zip
District   Available)

If no street address is available: _____   _____
                                                 (Township-Range)   (Section)

Checklist Preparer:

_____   _____
                        (Name / Title)   (Date)

_____   _____
                        (Organization)   (Phone)

_____   _____
                        (Street)   e-Mail

_____   _____   _____   _____
                        (City)   (State/Terr.)   (County)   (Zip+4)

Site Contact Info/Mailing Address: _____

CERCLA 105d Petition for Preliminary Assessment? _____   If Yes, Petition Date (mm/dd/yyyy): _____

RCRA Subtitle C Site Status: Is site in RCRA Info?   (Make selection)   If Yes, RCRA Info Handler ID #: _____

Ownership Type: _____   Additional RCRA Info ID #(s): _____

Site Type: _____   State ID #(s): _____

Site Sub-Type: _____   Other ID #(s): _____

Federal Facility? _____   Federal Facility Owner: _____

Formerly Used Defense Site (FUDS)? _____ If Yes, FF Docket Listing Date (mm/dd/yyyy)_____ Federal Facility Docket? Y N

Native American Interest?   (Make selection)   If Yes, list Tribe: _____

                                                   Additional Tribe (s): _____

                                                   _____

Site Description

Use this section to briefly describe site background and conditions if known or (easily) available, such as: operational history; physical setting and land use; site surface description, soils, geology and hydrogeology; source and waste characteristics; hazardous substances/contaminants of concern; historical releases, previous investigations and cleanup activities; previous regulatory actions, including permitting and enforcement actions; institutional controls; and community interest.

Geospatial Information

Latitude: _____   Longitude: _____

Decimal Degree North (e.g., 38.859156)   Decimal Degree West (e.g., 77.036783)

Provide 4 significant digits at a minimum, more if your collection method generates them.

Except for certain territories in the Pacific Ocean, all sites in U.S. states and territories are located within the northern and western hemispheres and will have a positive latitude sign and negative longitude sign. Coordinate signs displayed above are based on the State/Territory entry on page A-1. Geospatial data tips from the PCS Guidance document are available here.

PointDescription:Select the option below that best represents the site point for future reference and to distinguish it from any nearby sites. See additional information here.

☐ Geocoded (address-matched) Site Address
☐ Site Entrance (approximate center of curb-cut)
☐ Approximate Center of Site
☐ Other Distinguishing Site Feature (briefly describe):

_____

Point Collection Method: Check the method used to collect the coordinates above and enter the date of collection. See additional information here.

☐ Online Map Interpolation
☐ GPS (handheld, smartphone, other device or technology with accuracy range < 25 meters)
☐ GPS Other (accuracy range is ≥ 25 meters or unspecified)
☐ Address Matching: Urban
☐ Address Matching: Rural
☐ Other Method (briefly describe below):

_____

Collection Date (mm/dd/yyyy): _____

POINT-SELECTION CONSIDERATIONS

- Often the best point is a feature associated with the environmental release or that identifies the site visually.

- Use the curb cut of the entrance to the site if there is a clear primary entrance and it is a good identifier for the overall location.

- The approximate center of the site (a guess at the centroid) is useful for large-area sites or where there are no appropriate distinguishing features.

- Use the geocoded address if that is the only or best option available, but if possible use something more representative for sites larger than 50 acres.

Complete this checklist to help determine if a site should be added to the Superfund Active site inventory. See Section 3.6 of the PCS guidance for additional information.	YES	NO	Unknown
1. An initial search for the site in EPA's Superfund active, archive and non-site inventories should be performed prior to starting a PCS. Is this a new site that does not already exist in these site inventories?	☐	☐	☐
2. Is there evidence of an actual release or a potential to release?	☐	☐	☐
3. Are there possible targets that could be impacted by a release of contamination at the site?	☐	☐	☐
4. Is there documentation indicating that a target has been exposed to a hazardous substance released from the site?	☐	☐	☐
5. Is the release of a naturally occurring substance in its unaltered form, or is it altered solely through naturally occurring processes or phenomena, from a location where it is naturally found?	☐	☐	☐
6. Is the release from products which are part of the structure of, and result in exposure within, residential buildings or business or community structures?	☐	☐	☐
7. If there has been a release into a public or private drinking water supply, is it due to deterioration of the system through ordinary use?	☐	☐	☐
8. Are the hazardous substances possibly released at the site, or is the release itself, excluded from being addressed under CERCLA?	☐	☐	☐
9. Is the site being addressed under RCRA corrective action or by the Nuclear Regulatory Commission?	☐	☐	☐
10. Is another federal, state, tribe or local government environmental cleanup program other than site assessment actively involved with the site (e.g., state voluntary cleanup program)?	☐	☐	☐
11. Is there sufficient documentation or evidence that demonstrates there is no likelihood of a significant release that could cause adverse environmental or human health impacts?	☐	☐	☐
12. Are there other site-specific situations or factors that warrant further CERCLA remedial/integrated assessment or response?	☐	☐	☐

Preparer's Recommendation: ☐ Add site to the Superfund Active site inventory.

☐ Do not add site to the Superfund Active site inventory.

Please explain recommendation below:

PCS Summary and Decision Rationale

Use this section to summarize PCS findings and support the decision to add or not add the site to the Superfund active site inventory for further investigation. Information does not need to be specific but, where known, can include key factors such as source and waste characteristics (e.g., drums, contaminated soil); evidence of release or potential release; threatened targets (e.g., drinking water wells); key sampling results (if available); CERCLA eligibility; involvement of other cleanup programs; and other supporting factors. Attach additional pages as necessary.

Checklist Preparer Name   Checklist Preparer Organization   Date

Site Description
(Continued from page 2)

PCS Summary and Decision Rationale
(Continued from page 4)

Index

Printed in the United States
By Bookmasters